Reinventing Land Development: From Disdainable to Sustainable
5th Edition

Rick Harrison
Author

Adrienne Harrison
Co-Author

Prefurbia: Reinventing Land Development: From Disdainable to Sustainable, by Rick Harrison

Copyright © 2018 by Rick Harrison (1952-). All rights reserved. Except as permitted by the United States Copyright Act of 1976, no part of this publication may be reproduced or distributed in any form or by any means, or stored in a database or retrieval system, without the written permission of the author and publisher.

The word "Prefurbia" is a trademark of Rick Harrison.

Fifth Edition   Updated December 2018

ISBN-13: 978-0-578-41802-5 (perfect)

Information in this book is general in nature and should not be used without professional analysis and design and/or opinions and counsel of qualified consultants and attorneys. The publisher, and the author and contributors make no implied or expressed representation, warranty or guarantees of materials provided in this book and are not responsible for any errors or omissions or the results obtained from the use and/or implementation of such information.

Project Editor Editions 1 and 2: Greg Yoko
Associate Editor: Adrienne Harrison
Art Director: Char Thumser
Composition: Bob Buss
Cover Design: Rick Harrison

Forward

Why use Rick Harrison's Prefurbia methods?

As both a land developer and consultant to developers, I value Rick's intense focus on the efficiency and attractiveness of the land developments he has planned. He has produced neighborhoods that are innovative, highly marketable and cost effective. His background as a planner, land developer and civil software developer, is unique in the industry.

Connected Communities

Today our need for community is also stronger than ever. This desire, combined with an ongoing interest in fitness, has made walking trails one of the most popular amenities in new residential communities. Prefurbia designs connect the site with walking trails before roads are situated. This approach enables a safe separation of the two, promoting safe driving habits within the community.

Value creation

There are several factors that make traditional gridded developments age badly, but one of the most obvious is the fact that this style focuses driver's attention on the street and the seemingly unending garage grove effect along both sides. Prefurbia methods primarily focus the eye on the homes' front elevations and landscaping positioned along a meandering setback line. This creates a changing scale, yielding more value for the developer and the homeowner.

Efficiency

Rick's Prefurbia methods require less land devoted to street, to generate the same number of lots than traditional grid patterns - sometimes significantly less street area! This increase in efficiency means that there is less future street to be maintained (whether by the local municipality or the Homeowners Association), which can mean lower HOA dues for area residents.I strongly recommend that anyone who is considering letting their engineer do their land plan, for them to take a long look at the difference between the cookie-cutter subdivisions littering suburban America and what Prefurbia developments have to offer.

Skip Preble, President - Land Analytics, LLC

Table of Contents

About the Authors vi
Preface vii
Acknowledgements ix

SECTION 1: The Past and Present
Chapter 1: Typical Suburbia 3
Chapter 2: The Land Planner 19
Chapter 3: The Design 31

SECTION 2: The Present and Future
Chapter 4: Smart Growth and Green Issues 43
Chapter 5: Sustainable Development: A Practical Approach 57

SECTION 3: New Design Strategies
Chapter 6: Land Use and Environmental Conditions 67
Chapter 7: Transportation Systems 81
Chapter 8: Coving 87
Chapter 9: Mixed-Use and Multi-Family Housing 107

SECTION 4: Advancing Architecture, Urban Prefurbi & Regulations
Chapter 10: BayHomes 129
    Architectural Blending 138
    Architectural Shaping 144
Chapter 11 Prefurbia for Redevelopment 146
Chapter 12: A Model Coving Ordinance 153
Chapter 13. Technology and Education 165

Afterword 169
Terms and Definitions 173
Neighborhood Showcase 179
Author Biography 207
Photo Credits 210

About the Authors

Richard "Rick" Harrison

Rick is the founder and president of Rick Harrison Site Design Studio (www.RHSDplanning.com), a land planning & research company in Minneapolis, Minnesota, and Neighborhood Innovations, LLC (www.neighborhoodinnovations.com) which develops and distributes LandMentor land development technology and related training. The Design Studio was formed in the late 1980's (incorporated in 1993) to improve the quality of life through better neighborhood design. Rick's exceptional project approval record of 99 percent (when he conducts presentations), demonstrate his success with city councils, planning commissions, stakeholders, and citizens. His creation of the *Coving Planning Method* earned him the 1999 Professional Builder's Professional Achievement Award for Innovation in Land Planning. He has since introduced many more design concepts - all discussed in this book. Prior to forming Rick Harrison Site Design Studio, Rick was a land planner, land developer, surveying and engineering technician, and a civil engineering/surveying software developer (since 1976) as well. To our knowledge, no one has this unique blend of diverse experiences in the land development industry, and certainly none has fostered as many innovations in both land development design and software technology.

All of the plans Rick designs are to exacting precision, none are the rough sketches common in land planning. Rick also developed new software for the civil engineering and surveying industry. His patented LandMentor technology is the first system intended specifically for sustainable land development.

A more detailed biography of Rick can be found at the end of the book.

Richard L. Kronick contributed to much of the book including research and is a freelance writer and architectural historian. He has written extensively on Minnesota architecture, with more than 50 articles in publications including *Architecture Minnesota, Old House Journal, Progressive Architecture, the Journal of the Taliesin Fellowship,* and *Minneapolis St. Paul Magazine.* He has lectured on various architecture topics for the Chicago Art Institute, the Minneapolis Institute of Arts, the Minnesota Section of the American Institute of Architects, the Minnesota Society of Architectural Historians, and the Twin Cities Bungalow Club. Dick is a past president of the Minnesota Chapter of the Society of Architectural Historians and founded its newsletter, *With Respect to Architecture.*

Preface

If you are reading this book, you have an interest in land development. There is no single 'product' as expensive as the development of our cities. Yet, from an innovation perspective, land development has made little progress compared with the revolutionary advancement of most 'human-designed' products available today.

Over a quarter of a century ago, we noticed the nation was becoming more 'cookie-cutter' due to issues that plague growth and mindless software used to expedite plans. Rick Harrison Site Design was formed as a land planning research arm of a civil engineering and surveying software company I founded. The advantage of developing our own technology in-house enabled us to design neighborhoods with higher living standards, a preferred quality of life, yet used land and materials more efficiently. Thus, we coined the name Prefurbia.

The result was the creation of better neighborhoods, not just platting subdivisions *faster*. This book explains the problems of growth, then share with you market proven solutions successfully used on over 1,100 neighborhoods in 47 states and 18 countries.

Rick Harrison

Above: Example of a 'systems based' approach to site design, used in Avon, Indiana

Every one of the 1,100+ neighborhoods we designed gave us an opportunity to push the envelope further. A stepping stone to attain higher standards and striving for greater sustainability and affordability. What we have achieved so far is just the beginning. If everyone becomes more focused on serving the future residents goals above that of the municipality and developers, we can provide a preferred model of living - *Prefurbia*.

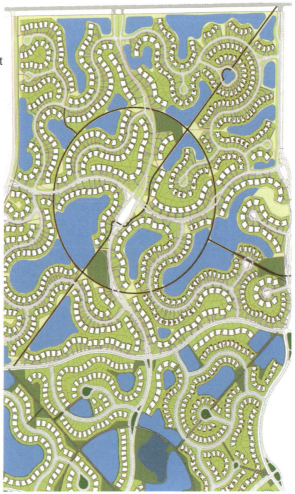

Just as a land surveyor describes every tract of land from the *Point of Beginning*, think of this book as the Point of Beginning for a new way to design, develop, regulate. Prefurbia precents a better way to build communities that serve as a model for a sustainable future, while also preserving the American dream of home ownership.

Acknowledgements

We are confident this book will influence positive change in the building and rebuilding of our cities. Over the years I was guilty of the majority of the problems described in this book.

I was born with a pencil in my hand. When others in grade school were drawing stick figures, I was figuring out how shadows and reflections worked. On the other hand, I paid little attention in school, sketching instead of passing exams. I was also fluent in trigonometry, the math used in software development and land calculations. Instead of a formal education, I embarked on learning on the job as an apprentice in land planning, civil engineering, land surveying, and in land development.

My success has been due to the following mentors (in order of date):

1968: Don C. Geake

If Don never hired me at 15-years-old, I would never have had the incredible opportunity to design neighborhoods. It would never have crossed my mind. I tell some revealing stories of my planning experience while working with Don. The situations that are still too common in the land planning industry.

1968: Calvin P. Hall

Cal took the time to mentor me in land planning while working for Don..

1970: Evelyn Wood Speed Reading

After graduating high school, I was functionally illiterate - again, concentrating only on art and math. Taking the Evelyn Wood Speed Reading course allowed me to read fast and with the comprehension required to educate myself by diving into multiple books as I pursued engineering, surveying, and other land related interests.

1970's: Abe Minowitz

My stepfather was one of the great industrialists of Detroit, Michigan. He gave me the opportunity to develop land by overseeing the completion of a New York multi-family development - at just 21 years old. This experience gave me the insight that if I were ever to be an effective land planner, I needed to gain a full knowledge of engineering and surveying. He was also instrumental in teaching me the "business" side of business.

1974: Paul Lederer

Paul hired me as a land surveying and civil engineering draftsman in Houston, Texas. He, and Chalmers Miller, a Civil Engineer, taught me surveying and engineering, from determining boundaries to designing sewers. Paul was a mentor that gave me the foundation to create precision designs. It was during this period that I developed a hobby of software writing to automate tedious computations, making my work easier. This was done using programmable calculators in the early 1970's.

1977: Bob Needham

Bob was head of the civil engineering section of Herman Blum Engineers in Dallas, one of Texas' largest consulting firms. I was passing through Dallas looking to get back into planning. Bob told me years later, that J. Stiles, their largest developer client, called one day and said: *if they did not get a planner on staff, he'd go elsewhere.* When Bob hung up the phone, it rang – it was me asking if they had any planning positions! I thus became head of planning for one of the largest consulting firms in Texas.

The first week I designed a curvaceous subdivision. I overheard the surveyors in a cubicle say, "*if he thinks we are staking out those curves, he's nuts -we are going to straighten out those streets.*" They did not know I had a full knowledge of engineering and surveying. I stayed the entire weekend – never left the office - using COGO (the standard software of the day) and punch cards, I did all of the geometry. That Monday morning I dumped the drawers of punch cards and hundreds of calculation sheets on the surveyor's desk, and said, "*Here it is – stake it out!*" I took much of my spare time and own income to invest in Hewlett Packard systems creating software to reduce time. Eventually, a surveying dealer sold 20 of my software packages, in the late 1970s.

1970's: Hewlett Packard

I worked within blocks of the main Neiman Marcus store. In those days, that store had the very latest land surveying technology. Why? Because the wife would shop for her clothes and the wealthy oilman could geek over the technology. One day I went to look at the latest, and I saw this incredible calculator being put on the display. It was the very first Hewlett Packard 41-C in the world for sale and it was to become an engineering standard for decades to come. n I began writing my first serious software on my HP-41-C. A few years later I sold my Honda motorcycle and bought a HP-85 and wrote my first commercially available software.

One day I received a call from the VP at Hewlett Packard Corporation asking me to write a surveying package for its new prototype desktop computer, the HP-87. Over the next two decades, we sold about $20 million in software systems to thousands of engineers and surveyors. The success from collaborating with HP allowed me to shift my focus to a new passion – improving the way suburbia was designed, and ultimately, to developing more sustainable land development methods.

Tres Lagos - McAllen, Texas - The largest master planned community in the Rio-Grande Valley.
Left to right. Rick Harrison, Mike Rhodes, and Nick Rhodes of Rhodes Development

Tres Lagos - McAllen, Texas - The main entrance to the 2,600 acre master planned community, a year after grand opening.

A resident testimonial (un-edited)

Good morning.

I live in Hunters Pass Estates. I have recently learned that your company was responsible for designing our neighborhood and just reviewed your website.

Speaking on behalf of our entire neighborhood we would like to thank you for the excellent design!
We are a very close knit neighborhood consisting of 39 homes at this time. The coving design provides each home with a view and also allows us to feel connected to each other due to the openness and ability to see other homes in the neighborhood.

Speaking on behalf of myself. My family had just built and were living in a home in a conventional neighborhood in St Michael. We never had the feeling of being part of our neighborhood. Our house was the second from the corner, when we looked out our window we could see 2 houses across the street and 2 on the opposite corner. The neighborhood had standard sidewalks but nobody ever seemed to use them much. Walks always seemed to be a planned event with the intended destination being back home. When encountering neighbors it tended to be a wave and friendly "hello" and you continued on your way around the blocks then back home.

My wife drove through Hunters Pass Estates when only the models existed. She called me right away and spoke not only of the models but of the neighborhood design. Needless to say, we were sold but had a hard time deciding which lot as each was unique but all offered the feeling of openness and the sense of connectivity we were looking for in a neighborhood.

We ultimately decided on a lot in the cul de sac Lydia Circle NE. . We couldn't have asked for more. I look out my front window or sit on my porch and can see the retention pond, lake, and 15 other homes in our neighborhood. The meandering sidewalks allow all neighbors to get to know each other better. Walks now tend to end up with our children playing with others, neighbors talking with neighbors, and more than one impromptu neighborhood get together that involved having pizza delivered to the house all of the neighbors seemed to end at that night.

The design also provides a sense of security. Our neighborhood has nearly 70 children with the majority being under 12. The coving design and narrow streets slows traffic, the wide meandering sidewalks provide a safe space for children to get back and forth, the openness allows you to see your children playing at other neighbors homes along with keeping an eye on each others property.

Your website, newsletters, and documents show your passion and commitment to the coving design. I Just wanted to thank you again and let you know the Hunters Pass neighbors are sold on the coving design. I'm sure I speak for many when I say we wouldn't move back to a neighborhood with conventional designs.

Please feel free to share any additional information that you think would be of interest to us on our neighborhood.

Brent Turner

SECTION ONE

The Past and Present

- Suburbia; a place where most new growth occurs
- The Land Planner
- The Design

CHAPTER ONE
Typical Suburbia

"A great city is not to be confused with a populous one."

— Aristotle circa 300 BC.

The land planning of neighborhoods impacts the success or failure of a region (i.e. sustainability). Suburban cities are comprised of a mosaic of 'profit driven' individual land developments haphazardly stitched together. They are rarely cohesive or functional.

Land developments designed in a few hours serve as a foundation that will exist for many decades, or more likely centuries. Homes may deteriorate, but they typically get remodeled, updated, or rebuilt upon the same block, lot, and street pattern that was originally designed.

There are few forms of design as permanent as the neighborhoods in which we live.

One size does not fit all

The approach most cities take in writing their land use regulations is summed up as "one size fits all." In this chapter we show how the people in charge of developing America's suburbs – city officials, civil engineers, planners, and land developers – start out with good intentions but end up instead with projects that increase housing and maintenance costs while diminishing the quality of life for residents, *while also harming the environment.*

Minimums that become standards

The typical suburban zoning ordinance begins with a purpose and intent statement. For example, here's one from a city near the Minneapolis-St. Paul, Minnesota, metropolitan area:

Purpose and intent

This ordinance is adopted for the purpose of:

- Implementing the approved comprehensive plan.
- Protecting the public health, safety, morals, comfort, convenience and general welfare.
- Facilitating adequate provisions for transportation, water, sewage, schools, parks and other public requirements.
- Balancing residential, commercial and industrial development and population to provide a tax base that can adequately supply the necessary level of services within the city.
- Providing convenient retail sales and service centers for residents.
- Facilitating continuation of commercial agriculture within the city.
- Minimizing conflicts between land used for agricultural production and land demanded for development.
- Conserving natural resources and maintaining a high standard of environmental quality.
- Conserving the natural, scenic beauty, rural character and attractiveness of the countryside.
- Providing for the administration of this ordinance.
- Defining the powers and duties of the administrative officers and bodies.
- Prescribing penalties for the violation of the provisions of this ordinance.

You will find similar verbiage in land-use ordinances throughout America. These are principles that any thoughtful person would agree with.

The problem is that in city after city, ordinance after ordinance, the thousands of words that follow those purpose statements do little to achieve the intent goals. In fact, most land use ordinances consist of nothing more than minimum lot dimensions and setbacks for single-family houses, multiple-family buildings, and commercial structures.
For example, in the same ordinance quoted above there exists a set of requirements for single-family housing as illustrated in Figure 1.1.

Area Requirements. The following minimum requirements shall be met for single-family residential development:

Minimum lot size	20,000 sq. ft.
Minimum lot width	100 feet
From arterial streets	100 feet
Front, from all other streets	30 feet
Side	15 feet
Rear	30 feet
Maximum building height	35 feet

Figure 1.1

Figure 1.2: No sense of space on the cluttered street in an upscale development in suburban St. Paul, Minnesota.

Suburban ordinances provide 'minimal' guidelines to control density of development. These regulations are intended to provide a certain provision or 'sense' of space.

But, as shown in Figure 1.2, there's little actual sense of space along most suburban residential streets. Why? Because those who are charged with implementing ordinances – elected officials, civil engineers, and city planners have little or no education in available methods and technologies specific to achieving an increased sense or 'preception' of space. This is one of the reasons this book was written - to provide both awareness of problems while introducing market proven solutions.

Those that see the word: *minimum* interpret it as: *requirement*. This encourages the land planner (designer) to line up houses like barracks: parallel to the street exactly at the stated regulatory 'minimum' setback distance to squeeze every possible home on the site.

There are three reasons why this situation happens:

- To maximize their client's (the developer) income it makes common sense for the *land planner* to place every home at the minimum allowable dimensions. They are striving for the greatest possible 'density' to maximize the number of housing units or commercial square footage that can be squeezed into a given parcel of land to provide their clients with more profit.
- City officials often agree because more homes increase tax income to support the city and fund operating expenses, as well as finance that new city hall, schools, library, and fire stations.
- Planning commission and council members assume that the dimensions allowed in the regulations will deliver a reasonable sense of space.

The result of designing to the minimums is the regimentation that people associate with today's cookie-cutter, monotonous, and mind-numbing American suburbs.

Land developers get the blame for our dysfunctional suburbia. However, the problem is caused by ordinances that state only minimum requirements and the developer maximizing density as profit criteria. The fact is the developer (almost) never designs their subdivision - it is those acting as the 'land planner' to blame. This problem is not unique to America, but in every country on our planet.

No requirement for creating neighborhoods with character
With these forces driving growth, how can we assure attractive neighborhoods will be built that add character and value to our cities as well as assure a great quality of life? The answer is: *we can't*.

Some municipalities have been successful in creating regulations that beautify. The city of Woodbury, Minnesota, enforces architectural controls for commercial construction. When driving along their arterial streets, passing the shops and stores, Woodbury appears more impressive than neighboring towns that have no such controls. Thus, Woodbury becomes 'the place to live' driving up both home and land values (and tax base). It is the rare city that has city staff talent and the will of Woodbury. Conversely many North Dakota cities during the oil driven construction boom feared that design controls would scare away builders. It will - *the shoddy and cheap cut-rate builders*.

Who actually designs most of America's new neighborhoods?
Much of our developed suburban landscape is laid out by civil engineering and land surveying firms, a breed that is notorious for their conservative approach.

When engineers and surveyors lay out developments, they think in numeric and linear terms: simple straight sewer systems, making that 10,000 square feet (minimum) lot exactly 10,000 square feet, not 10,006 square feet, and so on.

The typical engineer or surveyor rarely, if ever, considers homeowner's views when looking out their living room window onto a yard, or taking a safe and convenient stroll. They rarely if ever consider vehicular 'flow' or walking 'connectivity', or how to create a feeling of 'neighborliness'.

Why would they be concerned with such things? Even if the ordinance hints at these goals in the purpose and intent, it leaves no specific ways to achieve them. To make matters much worse, CAD (Computer Aided Design) software for laying out developments has always been focused on productivity, producing a plan as fast as possible. The term LPM, or lots per minute, is how many software vendors proudly market their products. One vendor boasts 250 LPM on their website! How much efficiency, function, and neighborhood character will those planning a development achieve at 4 lots per second? Meanwhile, the developer assumes that the people designing the subdivision design 'value' into the project, and do so within budget. Speaking of budget - many engineers charge a percentage of construction cost, thus being rewarded for creating the most construction infrastructure and the least profitable development - *i.e. unsustainable growth!*

Just bulldoze everything in sight
Three groups design suburban neighborhoods: the civil engineer, the land surveyor, and the architect all act as the 'land planner'. It is rare that anyone only does 'land planning' and nothing else for their income. Yet, land planning is what controls the success or failure of growth!

An 'architect' or a dedicated 'land planner' does not possess knowledge of civil engineering or land surveying. As illustrated in Figure 1.3, the easiest strategy for this type of land planner is to lay out a development with no thought about existing topography – the physical restrictions of the land itself. If the ground slopes the wrong way, no problem, the civil engineer will have to figure it out. What has been wrought by millions of years of geological evolution is quickly bulldozed.

First drafts of layouts for suburban developments, or 'conceptual plans', are created either just before or just after the purchase of the site. In a matter of a few hours – or in some cases with CAD, in minutes – the land planner sets lots and streets that will define millions of dollars of construction that are likely to exist for centuries. These quick 'sketch plans' are typically based on nothing more than a rough estimate of the outer perimeter of the tract along with regulatory minimum dimensional requirements for streets and lots. The term developers often ask: can you give me just a 'quick and dirty' layout? Often 'quick and dity' plans become *permenant unsustainable projects*.

Rarely does anyone consider *precise* topographic and vegetation conditions on the initial concept. Why? Because in many cases no such *precise information* is readily available and few developers spend the time and money to obtain it. Most in the industry think data from on-line 'interactive mapping' is accurate enough to design neighborhoods are fooling themselves. No design can be accurate without a proper precise site survey and topography done by a licesnsed land surveyor.

It's expensive to accurately locate the site boundary, every grade change, and significant tree. Thus, the land planners quick sketch based on 'bad data' becomes the basis for a series of important economic decisions that directly impacts the success of the developer. Can we afford the land? Can we achieve a profit? How many linear feet of street must be built? Earthwork costs?

Developers may abandon projects after relying on poorly thought-out land plans and bad data conclude that the cost of developing will be unprofitable. Or if the developer decides to go ahead and buy the land based on the quick sketch, they often discover too late that construction cost underestimated or the density had been wildly overestimated. The result is an unprofitable project with no "character building" traits because there was no money left to create them. Before the recession, land and housing values skyrocketed beyond reason. The developer could make all kinds of mistakes and still be somewhat profitable because of rising land values in a growing economy.

No street smarts

Because of the large lots and low density, Figure 1.4 appears to induce sprawl. However, this development is in a city without a sewer treatment plant, thus on-site septic systems and the associated land area required for the wastewater septic fields are unaviodable requirements.

Figure 1.3: A flat, boring, development in suburban Cincinnati, Ohio that destroyed the native habitat.

8   Prefurbia— Reinventing Land Development: From Disdainable to Sustainable

Figure 1.4: Large lot development creates longer streets per home as seen in this Minnesota development.

There are many reasons that justify what some consider to be sprawl. But less efficient infrastructure is not one of them. As you will learn in this book, suburban infrastructure design is generally more efficient than gridded cities of the past. The design methods of Prefurbia increases suburban efficiency a demonstrated average of 25% - in public street 'length'. Less infrastructure (street and utility mains) equates to more greenspace with less cost, and more 'design' opportunity.

When experts speak or write about how older cities did not sprawl like those being built today, they are correct. However, these experts leave out key points:

- The cities of yesteryear did not have today's restrictions and regulations that were enacted after environmental awareness began a half century ago. There were no 'wetland' laws, no slope restrictions, nor were there requirements to contain storm water on-site as is the case with most of todays developments.
- Older cities did not have the (absurdly large) setbacks sometimes required when transitioning land uses (zoning) abut each other on adjacent developments - simply to appease existing residents.
- Had the cities of the past been built with today's regulatory demands, they would have consumed more land - they would have *sprawled*. With the abundance of wetlands (previously known as 'swamps'), slopes, and required detention and retention ponding, the City of Minneapolis as an example, if built today, might have 'sprawled' double the land area!

One contributing factor to sprawl that can be dramatically improved through Prefurbia design methods is the design of local residential streets.

Figure 1.5: "Right of way" on a typical residential street.

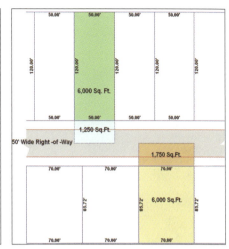
Figure 1.6: Typical suburban lots and their right of way segments.

Excessive streets are bad for (at least) three reasons:

- The most expensive item of a development is the public street. One mile of typical residential street costs about $3.2 million to build and much more for a city to maintain over its life span. This price varies around the nation, but the $3.2 million is close to a national average.
- A paved surface cannot absorb rain water increasing runoff (rainfall 'runs off' hard surfaces), contributing to flooding and the pollution of streams, lakes and watersheds.
- Homeowners pay for this excess. To explain why, you must understand 'right of way'.

Right of way defined

Most people think 'street = pavement.' However, the pavement is within a 'right of way', which is a corridor that envelopes the curbs, sidewalks, and utilitiy lines that contain sewer, water, gas, electric, and fiber optics. In American suburbs, the right of way dedicated to the municipality is from 50 to 66 feet wide. A privately owned lot extends to the right-of-way, not the curb.

Figures 1.5 and 1.6 show typical suburban lots. The homeowners of the green lot in figure 1.6 may think of their lot as 50 x 120 feet = 6,000 square feet. But they ultimately paid for half of the 50-foot-wide right of way (the people across the street pay for the other half). This is because the developer 'dedicated' the street right-of-way to the city factoring that land cost into the original lot price, thus cost of the home. So the actual land 'paid for' is 7,250 square feet. This cost continues as the home sells and resells.

If the town council prefers a more 'spacious look' from the street and therefore requires that 6,000-square foot lots be 70 feet wide, as with the tan lot in Figure 1.6, the homeowners' usable space remains the same, however, the added right of way length increases the total land consumed to 7,750 square feet. In other words, a 20-foot increase in lot width adds the cost of 500 additional square feet that the homeowner can't use or enjoy, yet they pay for it anyway!

Figure 1.7: The "garage grove" effect in a single family development in Fargo, North Dakota

Increasing lot width increases street length. The town council may have had the best of intentions, but by increasing lot width, they increase the costs of paving, utilities, street maintenance and run-off. Increased costs go directly to the house price which also increases taxes.

Land planners exacerbate this problem because they place structures (houses) parallel to and as close as possible to the street. This results in the longest possible street length to achieve density. Later in this book will be introduced more efficient and creative approach to design that reduces street length without sacrificing density!

The "garage grove" effect
Developers and builders must cater to today's multi-vehicle families who desire 2-, 3-, or 4-car garages with convenient access. A common way to provide extra car storage is to increase lot width and add garage stalls, which adds to driveway surface volume. Driveways are expensive - extra width increases costs and creates environmental havoc. Driveway costs are the responsibility of the home builder. More recently as suburban lots become narrower to achieve higher density, garages dominate the primary facade of a home. Combined, these factors create the "garage grove" effect. Driving down a typical development built in the past decade, less home and more garage doors are visible which becomes become the major (if not the only) architectural feature! Making matters worse is that garages seem to use the same white or sandstone steel door home after home (Figure 1.7).

It is the American love of freedom and luxury conveniences that today's cars provide – that influences much of our neighborhood designs.

It has become far too simple for cars to become the target for attacks on suburban sprawl.

One of the best-known commentators on automobile-driven planning is writer James Howard Kunstler. In *The Geography of Nowhere*, he says, "The amount of driving necessary to exist [in the current system] is stupendous and fantastically expensive…. The cost…in terms of pollution…is beyond calculation…The least understood cost – although probably the most keenly felt – has been

Figure 1.8: New Urbanist development in I'On, a suburb of Charleston, S.C., that squeezes everything together in an effort to reduce automobile use.

the sacrifice of a sense of place: the idea that people and things exist in some sort of continuity… that we know where we are."[1]

No doubt the automobile fosters some negative influence. However, the book's arguments written many decades ago, are different today as the cars are less polluting and far more efficient - as they will be decades from now. The ugly garage-grove look of development is solved through better design introduced later in this book.

Overdosing on cars has been made so exhaustively by Kunstler (and many others), we will not drag our readers through statistics to confirm the obvious: America is addicted to the automobile. Prefurbia embraces the automobile while softening its negative impacts.

The New Urbanists have responded to the automobile problem by suggesting that if they drastically shrink the scale of development it would reduce our dependence on cars and generate significant savings. For example, Andres Duany, Elizabeth Plater-Zyberk, and Jeff Speck, in "*Suburban Nation*" say, "…there has been no shortage of ideas designed to make the single-family house more affordable. The building industry and generations of architects have dedicated themselves to the task. The results – plastic plumbing, hollow doors, flimsy walls, vinyl cladding – are very clever, but all of them put together do not generate half the savings that can be achieved by allowing a family to own one car fewer."[2]

If that were actually the case – take for example the "I'On" New Urban development (a similar home to that on Figure 1.8). May 9, 2013: I'On Realty (www.ionrealty.com/find_home) shows the lowest priced home in the entire development as $448,500 . For this price the buyer gets only 1515 square feet, three bedrooms, 2 1/2 baths and apparently no garage at all. The same day – in the same region to I'On, is a development where you would find a DR Horton home and purchase the "Cedar" advertised for $373,900 with five bedrooms, three and a half baths, a huge usable front porch and a two car garage with 3,450 sqaure feet of living space - over twice that of the I'On home.

The purpose of cars
(Hint: not just transportation)

An affinity for convenient, personalized transport is nothing new. When European metalworkers learned to make spring steel in the second half of the 16th century, an important result was the development of reasonably comfortable carriages, as the preferred means of transport for wealthy people.[5] As with cars today, carriages instantly became status symbols. In Rome in the 1570s, it was said that two things were necessary for success: to love God and to own a carriage.[6]

And, as with cars today, 16th century carriages were not used solely to get from point A to point B. In fact, the availability of carriages almost immediately led to the popular practice of promenading: that is, driving up and down in a particular place at a particular time solely in order to see and be seen. By the early 17th century, every major European city had developed a purpose-built promenading street.

But the 16th and 17th century promenaders did not just see each other while coursing back and forth on the boulevard. They also communicated by voice and by passing notes to each other. In fact, the most important result of promenading was the arrangement of meetings – to conduct both business and pleasure.[7] Many of these meetings took place in the carriages.

Today, only two things have changed: the technology is more complicated and more levels of society participate in promenading. Americans want cars for the same reason people always have wanted personal transportation: because we conduct important parts of our lives in cars. Cars are part of our social fabric and we will have them or we will have something else that does what cars do for us. Public transport is not the same.

In other words, to understand why Americans tenaciously defend and protect their "love affair with the automobile," it may be more instructive to focus on the part about the love affair rather than on the part about the automobile.

Figure 1.9: Vehicle clutter.

The price difference is $74,600. Essentially, you have saved for your children's college fund by purchasing the D.R. Horton home vs. the I'On home - and doubled the living area!

This suggests that the scale of development is not the only issue that needs to be addressed.

If you prefer a more apocalyptic tinge to your theories, here is Kunstler's death knell for cars: "The Auto Age, as we have known it, will shortly come to an end…. We will almost surely have proportionately [fewer vehicles] per capita…. Possibly only the rich will be able to own cars."[3]

Yet, in the worst recession in modern history, car sales did quite well. The New Urbanism provides two ways to reduce reliance on cars. First, make suburbs look like inner cities by giving them drastically increased density to the point where daily destinations to the supermarket and the school are within walking distance of every home. Second, greatly beef up our emaciated public transport system so that vehicular trips don't require cars. Peter Calthorpe, a co-founder of New Urbanism, sums up these principles as follows: "…urbanism – defined by its diversity, pedestrian scale, public space and structure of bounded neighborhoods – should be applied

Figure 1.10: Boring, monotonous backs of houses exposed to the street.

throughout a metropolitan region regardless of location: in suburbs and new growth areas as well as within the city."[4]

This squeeze-the-cars-out attitude is where Prefurbia parts company with New Urbanists. It is a fantasy to assume that Americans will give up their vehicluar freedom - the ability to move at will in return for walking and its exposure to weather conditions that rarely provide days perfect for a stroll.

Whether or not they admit it, homebuyers base their evaluation of a neighborhood partly on the vehicles they see. A neighborhood of Mercedes and Cadillac cars signifies wealth, conversely a neighborhood of older rusted cars along the streets signifies blight. In larger suburban lot areas, where everyone is likely to have multiple stall garages, the impact of car clutter is reduced – but not eliminated. People protect their most expensive vehicles from weather and vandalism by sheltering them in garages. They tend to leave the least beautiful vehicles – older cars or teenagers' rust buckets, in their driveways, where everyone is forced look at them (Figure 1.9). Alleys and side - or rear access garages reduce the visual impact, but they don't solve the problem.

Showcasing bland rears of houses

Few houses have architectural detail on all four sides – not even those typically found in the New Urban projects. For costs, builders typically provide ornamental detail only on the front side of a house. The back and sides are typically left blank and unsightly (Figure 1.10).

Land planners too often think it's a good idea to turn the backs of houses toward arterial roads.

Multi-family housing used as a buffer between commercial development (above) and single-family housing (below)

Figure 1.11: Transitional zoning: townhomes placed as a buffer between detached houses and a shopping mall.

Thus, passing motorists get the worst view of a neighborhood. Noticing this, most officials choose one of two paths: Either they allow their towns to be ugly, or they require expensive buffers such as fences, walls, and berms between home and street.

Buffers become problems. Fences are not often maintained and age poorly. Homeowners may construct ragged, uncoordinated rickrack of fencing between their lots in a variety of materials, colors, and stages of decay. Earth berms with plantings may be better looking, but they consume expensive and excessive land and require costly long term maintenance. If the cost of screening devices went into architectural detailing, there may be no need for screening!

Figure 1.12: Transitional zoning, Multi-family units are at the entrance of this development in Hugo, MN, virtually hiding the higher-valued single-family homes.

Transitional zoning: hiding the expensive houses

In Europe throughout the Middle Ages and Renaissance, work was carried out in very small shops. But during the Industrial Revolution, from the end of the 18th century through the beginning of the 20th century, mechanization made it possible and economically advantageous to build large factories and thus employ more and more workers. As a result, work places became progressively bigger, noisier and smellier. In response, planners and city administrators came up with the concept of 'zoning'. They divided cities into separate zones restricted to heavy industry, light industry, retail, residences, and so on (Figure 1.11). They used multiple dwelling, light industrial and retail zones as buffers between upper-class residential neighborhoods and the most offensive industrial activities.

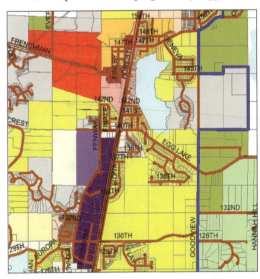

Figure 1.13: Zoning indicated by color in a Minnesota town. Most of the recent development has been based on the original mid-19th century government-mandated plat. For example, at right-center, the residential area (yellow) between Egg Lake [Road] and 136th Street has been shoe-horned into a half-mile square area, irrespective of topography and other factors.

The lowest-ranking members of society were usually stuck in close proximity to noxious factories. Thus, higher-ups insulated themselves not only from industry but also from "those people," whom they preferred not to see in their neighborhoods.

This practice – "transitional zoning"[8] – is still used and often for the same reasons (Figure 1.13). Citizens want to insulate and isolate their more upscale single-family houses from "those people". The usual excuse is to maintain property values, which there is some truth.

So how do those in control decide where to place the various zones? Usually, it is on the basis of previously existing land ownership boundaries. When Farmer Jones and Farmer Smith finally sell out to developers, the town 'comprehensive plan' likely designated the Jones farm as multi-family and the Smith farm as single-family housing. They often make such decisions even though a given configuration of land might not be the right size or in the best dimensions (or topography) for the use they have in mind.

They often decide to use the lowest-priced, highest-density housing as a buffer between the highest-priced homes and the most noxious land use. And since there is no noisy, smelly industry in most suburbs today, the most noxious land use is a high-traffic arterial highway at the edge of the development, or loading and waste disposal areas at the rear of commercial strip malls.

This common transitional zoning pattern has several negative effects:

- Passersby see only the lowest priced housing, thus cheapening opinion of the entire city.
- Passersby get the sense that the community is more crowded (dense) than it actually is.
- The majority of citizens are constantly exposed to most of the higher traffic and noise.
- Lower income families must worry about their children in proximity of high-speed roads.
- The view from apartment windows is either the highway, parking lots, loading docks, or trash bins of retail.
- Only the rich have the feeling and luxury of spaciousness.

Figure 1.14: These sidewalks (bright white strips), that even when houses are put in, will be of little use becaue they connect to no other walks.

All of this has an ironic boomerang effect that developers and community officials rarely consider: Potential home buyers never get to the hidden, higher-priced homes they seek because they are turned off by the lower-priced housing. This transition is especially common in "master-planned communities" – a marketing term intended to make very large subdivisions desirable.

Sidewalks to nowhere

In the southern United States, most subdivisions have sidewalks. But in the north, officials wrestle with whether or not to require them because citizens complain about increased maintenance of snow and ice-removal. Some cities require that developers install sidewalks when the streets are constructed. Then, when the houses are built, construction may destroy the sidewalks, essentially the sidewalks are then rebuilt and the cost is passed to homebuyers.

City officials often miss the most fundamental question:

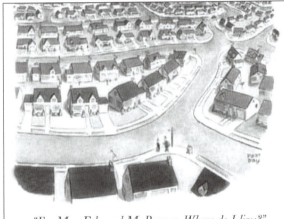

The cartoon below appeared in The New Yorker, 1954 and illustrates that suburban design has not been changed in almost six decades!

"I'm Mrs. Edward M. Barnes. Where do I live?"

Destination: Where can you walk to?

With the typical suburban pattern of large blocks, as shown in Figure 1.14, long winding residential streets that lead only to other residential streets, and zoning that places retail and work miles away, people are forced to drive wherever they need to go.
Therefore, sidewalks may have little purpose! In the end, the cost of sidewalks – which must include the cost of initial construction, ongoing maintenance, and stormwater management from increased water runoff – adds little real value, if the design has sidewalks as an afterthought!
Later in this book you will learn about entirely new ways to look at the design and implementation of both vehicular and pedestrian systems that add fuction, connectivity, safety, and efficiency while decreasing construction costs and environmental damage.

A more preferred way to develop (and redevelop) cities - Prefurbia.

References

[1] James Howard Kunstler: *"The Geography of Nowhere: The Rise and Decline of America's Man-made Landscape,"* Touchstone, 1993, page 118.

[2] Duany, Plater-Zyberk, and Speck; Ibid., page 57.

[3] Kunstler; Ibid., page 124.

[4] Calthorpe in *"The New Urbanism,"* Peter Katz, ed., McGraw-Hill, Inc., 1994, page xi.

[5] Cyril Stanley Smith and R.J. Forbes: *"Metallurgy and Assaying,"* in *"A History of Technology,"* edited by Charles Singer, E.J. Holmyard, A.R. Hall and Trevor I. Williams, Vol. III, 1957, Oxford University Press, pp. 34-37.

[6] Cardinal Charles Borromeo, in Jean Delumeau: *"Vie Economique et Sociale de Rome dans la Seconde Moitie du XVIe Siecle,"* who in turn is quoted in Mark Girouard's *"Cities & People,"* Yale University Press, 1985, pp. 118-124.

[7] Mark Girouard *"Cities & People,"* Yale University Press, 1985, pp. 118-124.

[8] In *The Language of Zoning, Planning Advisory Service Report No. 322*, the American Planning Association defines transitional zoning: "A permitted use or structure that by nature or level and scale of activity acts as a transition or buffer between two or more incompatible uses."

CHAPTER TWO
The Land Planner

"Form follows function – that has been misunderstood. Form and function should be one, joined in a spiritual union."

— Frank Lloyd Wright

As a career, land planning can be deeply rewarding. It feels good to drive through a neighborhood that you have designed. And it's a good thing there are intangible rewards because, according to the U.S. Bureau of Labor Statistics, salaries for planners are far from the top of the heap. In 2004, the median annual pay for urban planners (who, as explained later, may or may not also be land planners) was between $41,950 and $67,530. The highest-paid 10 percent earned more than $82,610 — about the same as an entry-level attorney.

Although a land planner's pay is not likely much, they shoulder great responsibility. For instance, a typical new suburban development our planning firm designs in 2018 contains 250 homes.[1] The U.S. national average new home price is (NAHB) $323,000. In other words, the average land development we design represents $81 million dollars of investment by home buyers.

And yet, using today's computer-aided design (CAD) systems and conventional cookie-cutter planning methods, the average 250-lot subdivision can easily be laid out in less than a day (one software vendor boasts on their web site they can layout those 250 lots in a singe minute!) If the planner is charging $80 per hour, that translates to $640.00 for the initial site layout — or about $2.50 per lot for the whole neighborhood. By comparison, the real estate agent who sells just one of the $323,000 homes will earn a commission over $22,000.

What is a land planner

Interestingly, the title "land planner" is not listed as a career option by the U.S. Bureau of Labor Statistics. This invisibility is reflected in local ordinances throughout America. In most U.S. states, there are no laws – or even rules of thumb – mandating particular levels of training, credentials, or expertise for anyone who plans the land, the people who lay out house lots, shopping centers, parks, public lands, streets, trails, and sidewalks.

Land planning is unique in this respect. By law in almost all 50 U.S. states, if a city is creating as little as a park plan (let alone an entire housing subdivision), the city must retain:

- A licensed land surveyor to determine the park boundary.
- A licensed civil engineer to create a drainage plan.
- A licensed landscape architect to choose plantings.

But the person who designs the *city* in which the park is located needs no license or registration whatsoever! All anyone has to do is put "Land Planner" on a business card to convince a developer or the typical city council that he or she is qualified to create a land plan — a plan that, once implemented, will probably exist for more than a century and will impact the lives of tens of thousands of people. The same city requires dogs to be licensed but land planners - not.

Furthermore, most of the people who design suburban developments today do so as part-time adjuncts to their primary lines of work. They are typically civil engineers, land surveyors, architects, landscape architects, landscaping contractors, builders, developers, environmentalists, or attorneys. We know of several instances when the land planning task was passed off on a junior drafter working in an engineering or surveying office because the firm's principal members couldn't be bothered with the task. Perhaps that's how your neighborhood was created!

The initial problem with land planning

Vague "concept" plans

The prevailing attitude in city government (and academia) is that deciding the precise locations of buildings, landscape elements, and infrastructure are best left to surveyors, civil engineers, and architects *after* an initial "concept" site plan has been accepted. As a result, planners are rarely held accountable for the accuracy of their work. Most planners feel no remorse about taking liberties in those concept drawings. This also fosters an adverse relationship from those that create vague approved plans and those that have to make the designs work.

One of the more common ways that some planners reduce the visual impact of large paved areas is by rendering lush greenery, but at the initial concept stage there is no guarantee that the

Chapter Two: The Land Planner 21

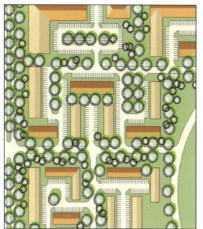

Figure 2.1: Drawing of a portion of a town center development in suburban Minneapolis, Minn.

Figure 2.2: Townhomes in Ramsey, Minn.

developer will actually install the landscaping the planner represents. A common tool of this illusion is the "tree stamp."[2] For example, it is common to see hundreds of green tree stamps hiding a large percentage of the massive paved area for a parking lot. But even if the parking lot does have trees, everyone who uses that lot will stand at ground level *under* the tree canopy where they will be visually assaulted by acres of asphalt and parked vehicles.

Figure 2.1 is a typical example. It is a concept drawing for a new town center in Minnesota.

Notice the profusion of tree stamps lining all streets. These tree stamps gives an impression of a lush, organic environment. In particular, note the many trees indicated around the townhomes. Figure 2.2 is a photo of a group of these same townhomes as built — with very little green space and few of the trees shown on the initial plan.

Figures 2.3 and 2.4 show another way planners might stretch the truth in a concept drawing. Figure 2.3 is part of a presentation drawing for a Minneapolis suburb. Note that the artist drew the picture from a vantage point that has a very long view through open space toward the housing. Figure 2.4 is a photo of the claustrophobic housing that was actually built. In reality, the artist's open space view is likely to only exist in the mind of the renderer not on the actual site.

Instead of searching for newer and more efficient design options that would increase quality space and livability, such as reducing the amount of pavement — it

Figure 2.3: A rendering provided by Doug DeHaan for illustration purposes, for this book.

Figure 2.4: The same development as actually built.

is easier for the planner to artistically de-emphasize it. City council or planning commissions who approve plans are not aware of this visual trickery.

Unfortunately, these practices are the norm — not the exception. Every day, city councils and planning commissions approve plans on the basis of these vague and often misleading concept drawings. This leads city officials and citizens to distrust developers, though the finger should be pointed at the land planner who knowingly duped the city and developer who paid them.

It is a reasonable expectation for planners to create accurate portraits of what they expect to be built. Tree stamps should be limited to the actual trees that are likely to be saved, and possibly a separate color used to represent trees that will be planted. Details such as driveways, patios, porches, and sidewalks should always be shown. Today we have better technologies such as LandMentor that allow us to show the intended finished look in any plan and in a 'virtual 3D setting'.

City councils and planning boards can make decisions based on reality rather than fantasy. And if citizens don't like the as-built look of a development that was presented honestly and approved by town leaders, those citizens will know that they should bring their grievances to the town leaders — not the developer. Today's virtual technology (VR headsets) eliminates falsehoods in planning.

Leave it for the next guy

When council or planning board members approve a vague concept drawing, they are passing on massive problems to the surveyor and civil engineer. If the planner has transferred hand-sketched boundary lines from a presentation drawing into a CAD system, it is unlikely that any of the lines will be accurate. Therefore, the surveyor or engineer will need to re-create everything. The entire site plan is dragged to the "trash can" and the surveyor or engineer, in anger, starts over from scratch. Typically, surveyors and engineers do not confer with the land planner, who they are already upset with for having to redo all of the work, and instead make revisions often destroying the design.

This means the job of re-designing a neighborhood is being thrown into the lap of someone who may not understand the goals of the original design and therefore unlikely to maintain its integrety. Yet this person makes revisions that will affect the bottom lines of the developer, builders, the city's maintenance costs, and the living quality of thousands of people who will eventually live in, work in, and visit the community.

Where the buck stops

Part of the blame for this situation can be laid at the feet of city officials. Most municipal mapping data provided to developers and planners does not contain precision. Thus, boundaries, flood limits, wetlands, and conflicts arising from easements are wrong — all of which are needed to correctly and accurately place lots, buildings, and streets. This is especially important with wetlands. Wetland boundaries are often drawn in city maps by draftsmen who trace vague information to create the city map. Wetlands are often quite different from what would be obvious to a wetland specialist standing on the site. Damage from this missinformation is huge

Often, another part of the problem is that the developer pressures the planner to begin designing when the developer has been unwilling to spend the money needed to obtain an accurate site land survey and associated data. Later, we provide specific solutions to these problems.

Who should do the planning

It may seem logical that a lot of land planning is done by civil engineers. After all, the civil engineers understand the technical aspects of land development, such as how to grade a site and how to lay out utilities. But that's just the problem; the typical engineer focuses only on technical details.

Few engineers think about maximizing views, increasing curb appeal, or how to the 'scale' of space or eliminate waste in design.

With existing planning (before Prefurbia), structures and streets parallel each other to achieve density. Thus, the planner consumes the site with streets in order to maximize the number of lots. But this typical scenario results in less usable lot area - and the most possible street length. If the consultant who engineers the development is being paid on a percentage of construction costs, they are rewarded for creating the most infrastructure. Should they reduce waste in the design, their income plummets!

Other land planners see themselves as social engineers. To name just two prominent historical examples, Sir Ebenezer Howard, the "father" of the Garden City movement in late 19th century England, and Robert Moses, often cited as the most powerful person in New York City during the mid-20th century, both felt they were improving society through planning. Neither achieved his goal. Many of today's best-known New Urbanists have similar lofty aspirations. Peter Calthorpe, for example, declares that diversity is one of the basic principles of urban design.[3] He believes that his efforts can create an environment that will bring together people of all incomes, ethnicities, and races at community focal points where everyone will exist in harmony. Yet, his organization the Congress of New Urbanism boasts of their 'gentrification' which pretty much eliminates such values!

Always ask the planner to detail their goals for the development, to determine if your visions are shared and to discover any unverbalized agenda.

About design charrettes[4]

Another common strategy used by planners is the charrette. A charrette is essentially an intense schedule of design activity wherein teams of designers work over a period of time to develop concept plans with input solicited from neighboring residents and city staff.

Life Experience

Rick gracefully declines

In 2005, a developer asked me to attend a meeting to discuss the creation of a major new 640-acre community in North Dakota. Arriving at the meeting, I found myself at a table with the developer, a two-person architectural team, and a market survey specialist hired to determine the types of commercial and residential buildings to be included. When the task of laying out the site came up, the architects said they wanted to take it on. They said that, in 1997, (with no prior planning experience) they had been hired to re-plan a downtown area that had been flooded. It had been so much fun that they now wanted to try planning this new development. The market surveyor said she likewise had never designed a plan before, but that she also wanted to be the land planner.

I was the only person at the table qualified and experienced to design the development. With everyone else's cards face-up on the table, I could see that my continued presence would only have led to a series of turf battles — literal and figurative. So I withdrew from the project and allowed the developer's "design team" to make decisions that would be critical to the success of his development.

I often encounter this situation: Developers step aside, allowing consultants of unsuitable expertise to run the show. It happens because, although the developer has been cast in the role of Moses to lead his people through the desert, the developer fails to take control of his team.

Fast forward to winter of 2013. The market survey woman whose employee told a North Dakota Mayor (who we were designing over 4,000 homes in his city) that 'coving', a design method explained in this book, would result in a development that could never be changed or redeveloped. We demanded what study or proof there was of this claim, and instead we and the Mayor recieved a written retraction of this damaging whim based statement.

Often, this is done to gain support for a proposal that differs from the existing ordinance. For example, the developer may want to build at a greater density than is currently allowed.

Many planners use the charrette as the mechanism to design a neighborhood. This form of design is inherently troublesome for the developer because it allows development opponents, typically opposing neighboring residents, to have an inappropriate level of input (power). In theory, the charrette can solve critical issues before submission, but that is highly unlikely, because the charrette process causes the following:

- The developer does not know who to blame for design changes that can destroy project feasibility. The intent of the charrette is to guide all those concerned to agree to a certain form of design – almost always a New Urbanist agenda. The developer assumes they have hired the planning firm for *their* expertise to design the development. When the charrette process is not used, a planner or team of planners, sits down and starts sketching based upon the desires of the developer, while trying to work within the regulations of the municipality.
- The charrette places many parties, all with their own agenda, in charge of design decisions. Few are likely to have the same goal(s) as the developer. Since the reason for the charrette is to obtain consensus by working through plans, the charrette may result in the developer catering more to neighbor concerns than that of the marketplace. In many instances, a charrette can be thought of as inviting the opposition to make critical planning decisions.
- A charrette is often a motivational tool designed to get people to agree to situations they may otherwise oppose. There is actually a certification for training at the National Charrette Institute (NCI) to learn how to influence a crowd through mechanisms that, if unchecked, may stretch the truth. The developer profits are likely to be sacrificed, yet, the planner can make hundreds of thousands of dollars in fees through out the charrette process!

Overreaching

With the combination of a winning personality and a talent for persuasion, a planner can be very successful.

Planners may accept assignments to design or consult beyond the scope of their knowledge and capabilities. Just because a planner is good at creating small subdivision plans does not mean they can design a large development. An expert golf course designer is not usually the right person to plan the adjacent housing development around their golf course, yet this is how things often work.

Life Experience

On the job training

During my six formative years in Don Geake's planning office, it was never suggested that I or anyone else in the firm should factor into our plans the cost of constructing the streets, sewers, and drainage systems we proposed. Furthermore, not once did we look at a topographic map to determine the best position for a street or home to minimize the impact of earthwork — either from an environmental or economic standpoint. That was seen as the job of the engineers and surveyors who would come after us.

Unfortunately, little has changed in the land planning field in the four decades since I worked for Don. Developers still assume that a planner takes engineering and surveying issues into account, but in most cases this is simply not the case.

Life Experience

My first and last charrette

A few years ago I was asked to work as a planning consultant for Seminole County, Florida. It was decided that I and several other experts would hold a charrette with the goal of convincing residents in an outlying area to hook up to existing sewer services and thereby allow an increase in housing density. When I arrived at the offices of the engineering firm who had subcontracted me, I met another design consultant (an expert on cluster planning flown in from Houston, Texas) who was just finishing his layouts on tracing paper over an aerial photograph of the area. I asked him what the scale of the aerial photo was so I could begin work on a site plan. He said he had no idea what the photo's scale was — yet he had used it as the basis for designing the entire area to be developed. It turned out that the squares he had drawn as homes were 300 feet by 300 feet!

Representatives from the county informed me that there were no mechanisms in place to guarantee what we proposed would actually be used, except to sway opinions.

In my presentation to locals the next morning, I prefaced my remarks with a statement that there was no guarantee that what we were proposing would actually be built — whether they approved it or not. After that, I promised myself that I would never again be a party to a misrepresentation in the planning process. It was the first and last time I took part in a charrette.

Layouts for multi-family and single-family housing should grow from very different strategies. If a plan shows attached housing laid out in a similar pattern as single-family, it's a sign the planner is over-extended.

Most of the elected and appointed officials responsible for approving a new development, members of the city council and planning board, typically have little or no experience in site planning. And yet, their role is combined judge and jury in what often looks just like a criminal court trial.

A city planner who is responsible for defining and enforcing regulations acts as a kind of prosecuting attorney. Since the city planner owes allegiance to the city, his or her emphasis will depend on which way the winds of power blow through city hall. Today, one of three major themes — environmental protection, social issues, or economic growth — tends to take precedence.

Meanwhile, whether justified or not, most developers petitioning to build in the city are viewed by the "judge and jury" as "the accused." Accordingly, the developer hires a land planner as "council for the defense." This land planner/defense attorney is beholden to the developer and therefore works toward the goal of increasing the developer's profit by maximizing the number of housing and commercial units on the site. Diametrically opposed to this, the city planner/prosecuting attorney is charged with upholding the city's rules and regulations, which always demand certain minimums for lot sizes, right of way dimensions, sanitation requirements, etc. Developer-defendants often choose not to attend such "trials," preferring to send their "mouthpieces" — the land planners — instead. Since land planners must function as defense attorneys, they need to have strong personalities and presentation skills to convince the "judge and jury." Unfortunately, some planners just sit and watch as good plans are shot down because they cannot deal with confrontation. And unfortunately, the developer also loses time and money which is ultimately passed on to the consumer in higher lot and home prices.

Master plans

A master plan is a layout for the overall utilization of a large area. Master plans typically show major land uses, defines densities, and suggests general traffic patterns. Methods and techniques for creating and displaying such plans have changed little since the 1960s. Often what passes for a "master plan" is nothing more than a very large housing subdivision. We can no longer afford to develop bedroom communities where shopping is 10 miles away and all residents need to commute an hour in each direction to work.

A master plan is often represented as a "bubble map," in which oblong circles indicate major land uses — housing, commercial, open space, etc. The scale is so large that no one can see what the final development is intended to look like. With today's technology and ability to generate detailed and accurate plans in reasonable time frames, we should strive to eliminate the use of these bubble maps. (They have some value as preliminary sketches used by planners to begin the design process.)

Essential technical knowledge that anyone planning land needs to know

As of the writing of this book, in the entire U.S., there seems to be no such thing as a university degree in sustainable 'suburban' land planning (the LandMentor System is essentially a complete course on suburban design with innovative methods). However, many schools offer "urban planning" degrees. An urban planner formulates plans for the short- and long-term growth strategy of a high density urban city. nThey study land use compatability, economic, environmental, and social trends. mWhen developing their plan for a community, urban planners consider a wide array of issues such as air pollution, traffic congestion, crime, land values, legislation and zoning codes. They focus on the macro issues in planning. Many students have opportunities to experience design on perhaps one or two small planning projects. But no program that we are aware of includes training at the micro level, in both the technical and community-oriented knowledge needed to transform a square mile of farmland into a functional neighborhood.

Essential *technical* knowledge needed by planners for successful sustainable land development planning:

- Determining precision boundaries (most on-line maps are not accurate)
- Drainage system design
- Sanitary system design
- Earth moving
- Construction costs

Determining precision boundaries

The process of land planning ultimately subdivides land defining legal ownership in the form of new lots. To accomplish this someone must first precisely determine the locations of all boundaries including those of wetlands and easements where building restrictions must be imposed. Until recently software systems that could be used to draw and simultaneously compute the precise coordinate geometry were tedious and required a background in civil engineering or land surveying. Land planners can learn the basics of surveying and civil engineering in order to produce accurate sketch plans the engineers need not revise. Prefurbia sets a new foundation for land development by utilizing new sustainable development technology that includes an education, to solve the above problems. We developed the LandMentor technology for this purpose which expands the knowledge introduced in this book and provides the tools for precision 2D and virtual design.

Drainage system design

Per code, every plan must include some system for capturing and diverting precipitation that falls on rooftops, walkways, driveways, and streets. If a planner chooses to bring in bulldozers to eradicate natural or existing drainage characteristics (it happens every day), expensive culverts, pipes, and other drainage structures then become necessary to carry away the runoff. This burdens a city with the cost of maintaining and eventually replacing all that infrastructure as it ages — infrastructure that might not have been needed if the planner simply used the site's natural drainage characteristics. The long-term costs to cities, developers, and homeowners are tremendous. Since rainfall cannot be predicted precisely, planners need to consider flood control into their designs. In some cases, it may be appropriate to base drainage calculations on a standard formula.[5] In other situations, modern computer modeling may be needed, software such as CivilGeo and others ahve been created for this purpose. In desert areas, planners must carefully consider "washes" — natural gullies that conduct flood water when it rains.

And as we all should have learned during hurricane Katrina in 2005 and Sandy in 2012, and later in Houston, planners working in tidal areas must consider hurricanes.[6] So why is it that engineers continue to construct millions of dollars in sewer pipes instead of using natural (cheap) surface flow? One reason is that a pipe-based network can be automatically designed with a few button presses in CAD - with little liability. A design utilizing surface flow would add more skill and hours to the design task. Compound that with the fact that engineering fees are charged by percentage of construction costs, the incentive to use surface flow in minimal.

Sanitary system design

Almost all traditional urban and most suburban sanitary sewer systems rely on gravity to maintain flow. For a gravity flow system to work, each successive pipe must go deeper than the pipe that feeds it. Deep pipes require deep, expensive trenches. Yet there is a limit to how deep trenches can reasonably be dug. When that limit is reached, an expensive lift station is needed to mechanically elevate sewage. Also, an expensive manhole is required wherever the piping changes direction. And

Life Experience

On the job training — NOT!

In Don C. Geake's office, we often kept developers temporarily happy by exaggerating the housing density that would fit in a given area. One way was to trace a site survey with a larger than normal scale. Another was to simply put fake dimensions on plans for streets and lots. While this made it appear that a large number of houses would fit on the site, the represented housing density was always lost when the engineer or surveyor tried to make the site plan work in the real world. The biggest problem was that, given an unworkable plan, the engineers and surveyors changed the plans to the point that they no longer functioned as originally planned.

Nearly half a century later, I still see plans that have been done this way.

when any of this equipment needs repair or replacement, the cost of tearing up streets and private property will likely exceed the initial construction cost.

However, new and improved technology is fostering the development of decentralized wastewater systems. Often, since many community centralized systems are already at capacity, these provide a sustainable alternative because they offer environmental and cost benefits. For instance, low-pressure sanitary systems are beginning to become an important alternative to gravity flow. In many cases, low-pressure systems cost less to install and maintain because they can be laid in shallower trenches and eliminate manholes and lift stations. However, these systems rely on individual pumps in homes and businesses, shifting the cost to home builders. That's why the decision to install such a system should be considered collaboratively by all stakeholders during the planning stage. Planners also should consider whether wetlands can be disturbed, and, if so, how much. Can wetlands be restored, maintained or upgraded and used as part of the drainage system? How much detention or retention of surface water is enough? This means that planners must collaborate with engineers at the initial stages of design.

Earth-moving

Planners should understand the methods and costs of moving earth and should factor these costs into all site plans. If a plan requires that large amounts of the surface be reshaped, the cost of every structure eventually built on that site will increase. Often with some forethought, the existing topography can be retained with benefits in terms of natural beauty and the cost of the drainage system. In later chapters, we present design strategies for managing topography and earth-moving.

Construction costs

Planners should know the cost ramifications of what they are proposing. Construction costs vary widely from region to region. In fact, two adjacent cities may have varying construction costs often due to differences in ordinances or union labor. It is impossible for a land planner to be an expert on everything, but it is possible to begin a new era of collaboration between engineering, planning, architecture and landscape artchitecture at the initial stages of design. Thus, collective skill and experiences are better than a draftsman in the corner cubicle determing the living standards of thousands of future residents and business owners

References

[1] The figure relating to an average development consisting of 250 homes is based upon Rick Harrison's personal business experience in 2006.

[2] Originally, tree stamps were actually rubber stamps that planners used to place images of large diameter tree-tops in hand-drawn birds-eye-view presentation drawings. Today the "tree stamps" are computer-generated, but the name has stuck.

[3] Peter Calthorpe: "The New Urbanism;" ed: Peter Katz, New York, 1994, p. xi.

[4] The process and the name charrette are legacies of the Ecole de Beaux Arts, the acclaimed 19th century Parisian architecture school. In the Ecole, professors regularly assigned design projects to their students with pressure-packed 24-hour or 48-hour deadlines. As a result, students often put finishing touches on their drawings as they were being driven to the final presentations in horse-drawn carts or, in French, charrettes.

[5] The standard formula was created by Robert Manning in the 19th century.

[6] In 2006, the Federal Emergency Management Agency (FEMA) and the Congress for the New Urbanism (CNU) were at loggerheads over the future of Biloxi, Miss., which had been devastated by Hurricane Katrina in 2005. FEMA had issued new regulations stating the height above sea level that buildings must be constructed in various parts of Biloxi. According to the regulations, in some areas near the ocean residents would need to place their rebuilt houses on 12-foot stilts. The CNU countered with the idea of "submersible" houses. Jim Barksdale, chairman of Mississippi's redevelopment commission, said the CNU plan was "just dead wrong" and a FEMA spokesman said, "Every time we get a big flood, we get people who say, 'We can build a flood-resistant house, which can get submersed and come out relatively damage free.' But the economic damage to that building may not be lessened very much, because the contents are damaged, the drywall has to come out, the electric's gone." "Battle for Biloxi," by Jim Lewis: New York Times Magazine, May 21, 2006.

CHAPTER THREE
The Design

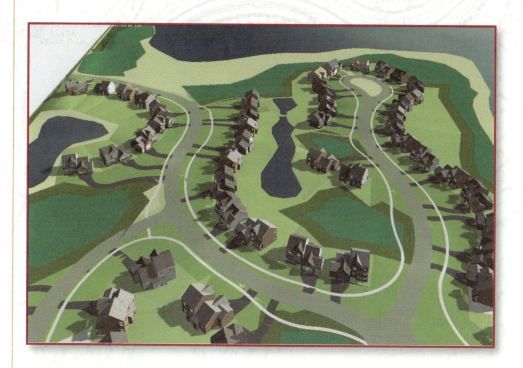

"The subdivisions of suburbia are conceived as shopping centers for housing and only later (if at all) as communities."

— Andres Duany

Designing a grid (or variations of a grid) is the most common pattern for a subdivision. It is incredibly easy - simply use the "offset" command in a CAD package. Free-form 'organic' design can be tedious when all components of the design, such as streets, setbacks, sidewalks, form their own pattern. CAD software makes a mindless design situation worse by providing only the automation of known processes that have been around since the days of horse and buggy. By relying on these automated features, entire sites can be accurately designed to surveying standards in short order. But are these quickly produced subdivisions desirable?

Prefurbia methods are mathematically challenging. There is no 'software button' to press that automates Prefurbia. When it comes to creating a neighborhood that contains well balanced elements, there are no short-cuts. Design professionals cannot rely on an automated software feature to do the thinking for them. They must instead take the effort to create exciting, efficient, affordable and functional neighborhoods that build character that can last for centuries. *There is no automatic function for sustainable growth!*

The real costs of simplistic planning

As was explained at the beginning of Chapter 2, an average new suburban development of 250 homes represents approximately $81 million dollars in investment by the initial home buyers.

And yet, as we explained, using automation to produce conventional cookie-cutter planning methods, the gridded 250-lot subdivision can easily be laid out for about $640 in planning fees.

Statistically a family will live in each home only six years. The average family size is 3.14 according to the 2000 U.S. Census. Over a century, the 250-lot neighborhood will house roughly 13,083 people (100 years divided by 6 years for each family multiplied by 3.14 [for the average family size] multiplied by 250 homes = 13,083 people).

This means during the next century this neighborhood will cost just under 5 cents per person to design in today's dollars ($640 for CAD design divided by the 13,083 residents = $.049)

A minimums-based process

Typical suburban ordinances are written with the "minimum" dimensional control looking like this:

R1 Zoning	Minimums
Lot Size	10,000 sq.ft.
Width at Front Setback	80 feet
Front Yard	30 feet
Side Yard	10 feet
Rear Yard	30 feet

Land development is a business – and like any business there is a wide variation on the personalities of those that own and manage that particular business.

Take a look at how two very different developers would utilize a minimums based system on a single family development:

A tale of two developers (using minimums-based ordinances)

Both developers had been using the same local engineering firm that designed most of the land developments in town. The engineers in the firm always follow the minimum standards – they never question any regulations, they rarely rock the boat. These engineers are numbers people, thus, a 10 percent park dedication will be exactly 10 percent, no less and no more. The 80-foot minimum width of lots will all be exactly 80.000 feet along the front setback, which will always be 25.000 feet from the street right of way.

Developer "A" (Mike) is known for shoddy cut rate projects. Mike has been an embarrassment to the community and those on the city council shudder every time they see one of his submittals, knowing the homes will be built with the minimum of landscaping and architecture. Mike is frugal. His accounting background works well with the engineering firm – they are numbers guys. Mike has a sign behind his desk that reads: "A penny saved is a penny earned." Mike never gets in front of the planning board and he never approaches the neighbors before meetings to gain support, so the neighbors assume what will be built will negatively affect their home values. As a result, the neighbors organize and show up in full force to fight his subdivisions. When the engineer gets up to present the plan – only the mathematics are discussed:

Figure 3.1: Developer A's grid subdivision (bottom half) vs. Developer B's more creative approach (top half) – Coachella, California.

"Uh, as you can see our plan – Lost Oaks - has uh 240 single-family lots as per ordinance 2.1.5 as per subdivision land regulation 3.6.10, subsection 5-A. Uh, of the 240 single-family lots, all meet the minimum 10,200 square-foot minimum…"

This mundane presentation goes on for 20 minutes, boring the council members who may daydream and drift off elsewhere. The city council asks if anyone is there to comment. After six grueling hours of neighbors attacking the "project" on how it will ruin their home values, the council votes at 2AM. They vote "yes" because it meets all of the citys regulations, but they all make comments on how they wish the plan could be better. And, of course, they blame Mike the developer.

Developer A's unsightly rear yards look best the day they are built and will clutter up with 'stuff' over time.

The reality is, the blame should be on the planning commission, the city council, the consultants, the mayor, the administrator. They accepted a system from their planning director that does not promote the best development within economic reason for all sites submitted.

Mike really does want to do a good job. He wants to be appreciated but does not understand why he is always seen in such a negative light for providing the much needed affordable housing. His engineering firm works by the numbers as well. In all likelihood, Mike hired them because they get the approvals and (like most consultants) tout 'sustainable' on their website. They simply take direction from their client and do not offer ways to make designs better. Mike's engineers feel it is not their job to change Mike's opinion and not worth the risk to introduce new concepts. What if their ideas don't work? Then, they will likely receive the blame and lose a valued client.

Mike thinks he must be doing something right, because his revenues will be several million dollars a year. Anyway, why give Mike such a hard time? He is the only one supplying affordable housing in town. He does not see anyone else stepping up to the plate to provide these homes.

Developer "B" (let's call him Joe) is a member of the Chamber of Commerce and heads up the local Sierra Club chapter. Joe looks at every development as part of his legacy of making a positive impact on the growth of the town he was raised in, and yes, he does still live in. The homes that Joe builds have nice curb appeal and landscaping details. Joe makes sure that the neighborhoods contain walks and social gathering places, even though it is not in any ordinance to require them.

Joe works under the same ordinance with the same regulations as Mike, but he feels that he must guide the engineers in town to be sure that the 10 percent open space is actually useable, with walks that lead to the space. He also advises the engineers that maybe the density is a bit too much, explaining that dropping a few of the allowed homes can make the neighborhood a bit more attractive. His neighborhood will look and feel different from others. It will be inviting.

Developer A's garage-front homes that lack curb appeal.

Before any meetings, Joe personally visits with the residents by holding two catered workshops at different times of the day so that most can attend, convenient to their schedule. The presentation explains all the benefits, the architectural control and the value that this neighborhood will add to the city as well as the neighbors property values. Many still are not in favor of having those homes in their back yards, but know overall it could be far worse - Mike could have been the developer.

At the public meeting, the engineer sits in the back and Joe gets up to the podium...

"Tonight we will present Preservation Oaks, a new neighborhood that will be a place that residents of our town will enjoy, not just today, but for decades and, hopefully, centuries to come. Our landmark community spaces where neighbors will congregate create harmony, as you can see on the screen, with our three dimensional animation and for the council through VR headsets that place you in the site..."

This goes on for 20 minutes. At no time did Joe mention a single dimension, volume, or any other defining number – nor did the engineer speak about, well, engineering stuff.

After 20 minutes of comments from the opposing neighbors, the city council votes an enthusiastic "YES!" No one really cared that the housing Joe built due to the extra architectural control and site amenities, would no longer be offered at prices the average family could afford.

From an actual "plat" and numbers standpoint, the engineering varies little between the two developer clients... they have the same basic density, same expenses, same regulations, and in the end the engineer makes the same amount of money.

There is generally no incentive for any developer to go beyond the absolute minimums. Developer "A", Mike, wants to do better, but cannot see a profitable reason given the city's codes.

Remington Coves, Otsego, Minnesota. An aesthetically pleasing (Prefurbia) site plan, with meandering centrally located open spaces and great walking connectivity.

The minimums-based system

As explained in the last example, zoning regulations and design ordinances are based upon minimums. Therefore, they are guidelines that can be improved upon by exceeding them. However, also acknowledge your strengths and weaknesses in regards to public speaking before presenting.

Understanding the approval process

The first rule of public speaking is to be prepared. The second rule is to know your audience. Does your development approval hinge on the vote of a planning commission or city council?

The system unique to the United States is one where local citizens decide the fate of the development being proposed, often, even if it meets regulations. The voters are people who want to serve on boards like the planning commission and council for a variety of reasons, mostly because they are concerned with the growth and management of the town their families live within. Some may serve because the power of the position is enticing. Some will have a personal agenda. Assuming the people serving are good people, overall, this flawed system works somewhat well.

Meetings involving growth issues (planning and zoning) are directly concerned with the developer's designs. The city council, however, deals with all issues concerning the town, police, schools, sewers, the "Potato Day Parade," etc. And, they often have the final vote on the developer's plan, yet have so little time to grasp the benefits (if any) of the development.

By contrast, the Planning Commission is advisory and they typically cannot give the final "yes" or "no." If the developer loses the Planning Commission's vote, they can still get the city council to vote yes. If the council gives a "yes" vote, the development will be built.

Developers, land surveyors, architects, planners, etc., usually concentrate their meeting efforts with the city planner. The city planner may be employed by the town – or an outside consultant that is hired to represent the town's interest. The city planner may have his own agenda. As an example, they may favor only New Urbanism and only promote that very 'personal' agenda.

It is essential that the developer and his design team focus on those that have the authority to vote for or against the project. This means the teacher, accountant, retiree, banker, and other common folk serving on the council. Do they want to hear engineering or zoning data? No! They want to learn how this development will be good for their community, and the residents who will reside within its borders. Therefore, the developer and designer should make sure that their neighborhood will actually be good for the community, and then know how to sell it! The council must feel as if they could live in the neighborhood.

Why PUDs may not be the answer

A PUD (Planned Unit Development) ordinance is written to give developers and planners flexibility in design, specifically to be creative. PUD presentations are often infomercials on the exciting new development, complete with earth-tone renderings and pictures of the utopian living offered by the special neighborhood being presented that night.

As stated in Chapter 2, the rendering may misrepresent what the actual development will look like on the ground. After the PUD is constructed, the council members may visit a jungle of concrete and rooftop (in areas they thought were to be green) and start to question the concept of having a PUD ordinance at all.

There are often many vague areas of the PUD ordinances that frequently get left up to interpretation. For example the PUD could promote 'architectural character', a non-specific term that is left to interpretation. Another problem of PUD is that a great presenter can get a mediocre plan approved more easily than a great plan introduced by a mediocre presenter. A clear set of rules based upon rewarding density for going the extra effort would solve many problems.

Why "points system" or "forms based" ordinances are not an answer

In many towns, planners may offer a newer form of ordinance where the developer earns "points" based upon the design and features of the plan. At first this seems like a better choice, but iis not.

Like the "minimums" based system, "points" are set by a round table discussion. George, a banker, insists that 10 points should be awarded for having speed bumps – Beth, a florist, insists on eight points for a playground with three swing sets, etc. The points system can sometimes become so complicated that people lose focus of their goals. While well-intentioned, these numeric based ordinances are likely to be worse than the minimums, a PUD, or no specific rules at all.

This brings us to the latest entry in the attempt to solve regulatory problems, the "forms-based ordinance" which replaces minimums with relationships of how far buildings must be positioned to other buildings or infrastructure. This ordinance pretty much guarantees monotony, it's a regulatory system that controls a rigid relationship between structures and streets. And again, the relationship of forms based systems introduces complexity to the process that could cause confusion to those that must vote YES for that development to become a reality. Keep in mind, the more complex the regulations, the more reliance on the planning consultant, assuring their job security and increasing their billable hours.

Breaking the minimums

The minimums-based system, for all of its faults, does actually work. It's pretty cut and dried – build this size or you are not likely to get approved.

The only case where the planner can easily justify breaking the rules is when the intent of the ordinance is exceeded. If the proposed development is an outstanding plan that assures a high standard of living and will become an asset to the community, that development has a good chance of being approved - even if some of the minimums are not met. For example, if the minimum lot size is 10,000 square feet, and the developer offers an alternate plan that is far superior in design to that which the ordinance minimums would allow, but is asking instead for a *minimum* of 7,000 square feet, with an *average* of 11,000 square feet, it is likely to get a "yes" vote. That is right – by holding and exceeding the "intent of the ordinance", the reasonable citizens that serve on planning commissions and city councils will likely go against a minimum regulation for the good of the citizens. The problem here is that the planner may not possess the necessary skills or experience to convey this idea. New rules are needed that can make everyone better stewards of our growth.

A "win-win" ordinance

The modern ordinance should be written to assure a minimum design standard, but even more importantly, one that would reward those going beyond the minimums. With a good set of rules that is easily understood, Developer "A" types can be transformed into Developer "B" types.

Reducing barriers to good design is an important goal. As an example, there is an ordinance that pertains specifically to the design method of coving. (Coving is explained in Chapter 8 and a sample ordinance is found in Chapter 11.) The principles in this ordinance rewarding better development could be applied to all forms of regulations.

A look at the spacious inviting park-like front yard setting in Sundance Village in Dickinson, North Dakota.

Coved design is about the efficienct use of a site: to design a development that creates less environmental impact and is less costly to develop than conventional development, yet looks and feels far more spacious and luxurious. Reduced costs could go directly into the developer's pocket, but a well written ordinance will encourage that developer to use the money saved in construction towards community building features which eventually should make the developer more profit through increased premiums and expedite sales.

Holding the *original intent* of the municipality is critical. Where do the physical "numbers" for the minimums come from? Some municipalities (or consultants) simply mimic that of other towns, while others start from scratch with informational workshops. At the end of the day, many of the dimensions boil down to emotions of council members. For example, council members Tom, Carol, Mike and Angie eventually may agree that a 70-feet-wide lot is too small, but they would accept 80-feet as a minimum. This is not based on anything but what they 'feel' is right.

One size does not fit all

The perception of a minimum lot size is regional. A 6,000 square feet lot would be considered far too small in many areas of the Midwest, yet that same 6,000 square feet might be considered a large lot in the south.

A side benefit of properly executed 'coved' design is a significant increase in average lot size. If the local minimum is 10,000 square feet, it would be typical for a coved neighborhood to have a 14,000 square foot average. Yet, the street length to achieve similar density could easily be 30% less. The benefit to both developer and municipality would be a large reduction of public street length at densities similar to a conventional plan following the same rules. However, for the developer, there is likely to be an offset by other expenses like increased manholes, sod surface, more driveway surface (about 30 percent additional), and if there is rear yard screening, increased fences or walls due to the larger average rear yard space provided by coved plans.

Unless the larger average lot is used to build a correspondingly larger home (unlikely), it may result in excessive space. This is counter to Smart Growth principles. It makes more sense to use the minimum lot size as an *average*, allowing for a minimum *coved* lot to be about 25 percent less than the current ordinance minimum. The exception to this rule may be if the original minimum lot size is already small.

From subdivision to sustainable land development

The ultimate solution is to develop an incentive-based ordinance, not a minimums-based ordinance, encouraging all developers to create functional neighborhoods with character.

As a municipality, the staff, administration, councils and public should determine the qualities that they preceive will build their community's character. A front porch, fountains, picket fences, and tree lined streets may make sense in Boston, but in Santa Fe with the Adobe architecture and shortage of water, it may not be a fit.

The incentive-based ordinance leads all developers to a path that will help them infuse character into the development, while providing benefits to the ones that matter most, the home buyers and residents in the community.

Add to the above idea function, efficiency, connectivity, flow, embracing natural alternatives, conserving energy, increasing open space, maintaining affordability, etc, and you ultimately create a more sustainable quality of life.

We call this idea, Prefurbia - a preferred way of living.

SECTION TWO

The Present and Future

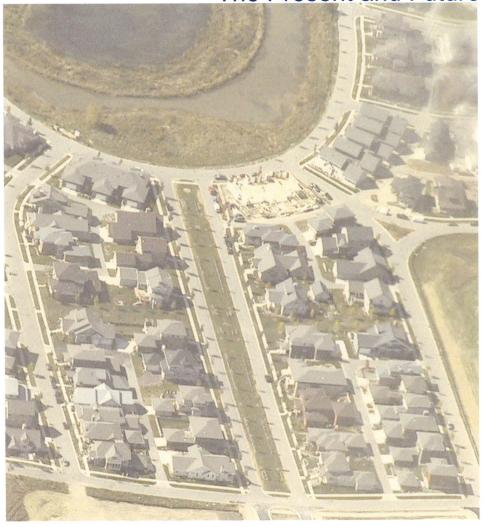

- Smart Growth and Green Building Issues
- Sustainable Development

CHAPTER FOUR
Smart Growth and Green Building

"Future economic prosperity depends on building a new, stronger foundation and recapturing the spirit of innovation. Innovation has been essential to our prosperity in the past, and it will be essential to our prosperity in the future."

— U.S. President Barak Obama

Smart Growth

Smart Growth as defined by the Environmental Protection Agency:

EPA: Smart Growth Principles

Based on the experience of communities around the nation that have used smart growth approaches to create and maintain great neighborhoods, the Smart Growth Network developed a set of ten basic principles:

- *Mix land uses*
- *Take advantage of compact building design*
- *Create a range of housing opportunities and choices*
- *Create walkable neighborhoods*
- *Foster distinctive, attractive communities with a strong sense of place*
- *Preserve open space, farmland, natural beauty, and critical environmental areas*
- *Strengthen and direct development towards existing communities*
- *Provide a variety of transportation choices*
- *Make development decisions predictable, fair, and cost effective*
- *Encourage community & stakeholder collaboration in development decisions*

All of the above are honorable goals, but, are they sustainable? If only a few of the above principles are applied to a project, is it still considered 'Smart Growth'? A typical development or redevelopment project cannot always satisfy all of the 10 principles. For example, not every urban redevelopment would contain multiple housing choices which implies housing price point variation, or have ready access to mass transit.

Take a look how strategies in Prefurbia satisfies (or not) the 10 principles of Smart Growth:

Mix Land Uses - *inclusive or exclusive*?

In dense urban environments, intermixing commercial and residential uses (often within the same building) is not unusual. Do all urban residents enjoy the extra crowds and noise of adjacent restaurants and shops? Gentrified (upscale) urban neighborhoods are often popular destination places for a larger regional population. For example, suburban Chicago residents often drive into the city core to enjoy the vibrant urban night life, then return to their quaint, quiet (and safe) space in the suburbs. It is doubtful that the same suburbanites would flock to downtrodden high density areas of Chicago that also provide 'walkable' restaurants and shopping.

Intermixing uses is a desirable goal, but businesses must also attract enough customers to be profitable. Few developers have the diverse experience or knowledge to successfully implement a mixed-use development in the suburbs.

A developer's 'comfort zone' is either in residential or commercial uses. These are two very different markets. Many developers are also builders, especially in the commercial market. A home builder will often view adjacent commercial as a negative factor to selling their suburban homes. The residential builder buying 100 acres for lower density suburban housing is not likely to see how placing a high density mixed-use adjacent to their spacious lots as something that will lead to increased home values and faster sales. However, intermixing residential and commercial such as the Neighborhood Market Place (chapter 9), provides a viable suburban solution for both residential and commercial developers.

Individuals sitting on suburban planning commissions and councils are not likely to be easily sold on approving high density mixed-use development in their quiet little haven. After all, they did not move out 'there' to live in a crowded and noisy city.

Suburban councils and planning commission members are savvy enough to know that 'Smart Growth' presentations showing well-dressed people within dense urban spaces are certainly are not

An architecturally attractive TND Town Square — Liberty on the Lake, Stillwater, Minnesota.

the residents of 'Mapleville', an area of casual dwellers. The urban 'image' they likely relate to are of decaying areas that followed the same tight grid design model of 'Smart Growth'!

Prefurbia design techniques make it possible to deliver a suburban sense of space at more aggressive densities, with strategically placed retail and professional services. With Prefurbia a balance can be created that supports a successful business atmosphere within a stroll of homes.

Chapter 9 explains how this can be accomplished.

Compact Building Design

A common 'Smart Growth' solution is to squeeze smaller homes on miniscule lots. The average home size according to the NAHB (National Association of Home Builders) varies each year, but it typically hovers around 2,500 square feet. This creates 2,500 square feet to heat and cool. If home size is reduced, it will reduce energy consumption. A 20 percent reduction in size would have a direct 20 percent reduction of energy use. A good architect can design a home with better 'curb appeal'; however, a great architect can make a 2,000-square foot-home 'feel' like 2,500 square feet

Prefurbia techniques position homes compact enough to be marketable in a manner that also delivers a "feeling" of less density - thus, increased space.

Ask 100 home owners how many square feet their home is, and most will give a close answer. Ask the same group how many square feet their lot size is and few are likely to know. This is an advantage for Prefurbia planning, because a 5,000 square foot lot can be designed to 'feel' as if it were 9,000-square feet. With an increased perception of space, higher density is easily justified.

A 'Smart Growth' lot is limited to a narrow yet deep rectangular shape with the home being correspondingly narrow and deep. With this configuration window locations with quality views is limited compared to a home on a wider lot of lesser density. The majority of the exposed exterior wall surface parallels the neighboring side yard - viewing directly into their house. 'Smart growth' homes are positioned close to each other. Side yard windows are placed to let light in and perhaps emergency access, but they cannot open up views from within the home unless it is on a corner lot.

Typical urban older neighborhood in Minneapolis, Minnesota (note: the 'wetlands' were bulldozed over back then).

A narrow home in Prefurbia will either be angled to the next home or staggered, allowing panoramic views from within the home even if placed within interior lots. A Prefurbia home can be "shaped" to fit an irregular lot providing a market advantage for the builder that would be impossible to replicate on a rectangular lot (as explained in Chapter 10).

In theory, compact buildings reduce energy consumption. If an architectural floor plan is compact yet has excessive waste in terms of space (large percentage of floor area in halls, stairways, utility corridors, all visual blocks to sense space, that must also be cooled and heated), then efficiency suffers. Again, skilled architects can create space with less waste, making a well designed 2,000-sqaure foot home function just as well as a poorly designed 2,500 square foot home. If a large percentage of the home is being consumed by wasted space, it is time to seek out better design.

With Prefurbia, window placement, room functions and floor plan are often coordinated components of the overall neighborhood design, improving the quality of life and the all important 'market edge'. Only when both architectural space and neighborhood design are integrated can a development offer compact design where the resident will not 'feel' compressed space. Interior space becomes a major component of the overall neighborhood function in Prefurbia planning.

Create a Range of Choices

Most developments designed at our studio provide a wide range of housing choices. Still, quite a few of our neighborhoods have similar-sized homes, on similar-sized lots, at similar price points. 'Smart Growth' developers often desire a similar range of home choices. However, most developers prefer to concentrate on a singular price point.

Both Prefurbia and Smart Growth developers are more likely to offer more diverse housing choices than the typical conventional subdivision-oriented developer.

A major difference from Smart Growth (i.e. New Urbanism) in actual implementation vs. theory, is few (if any) Smart Growth neighborhoods are affordable. Most Prefurbia neighborhoods are in the lower to middle income range, serving the mass market, thus the 'greater good'.

More detailed information about range of choices can be found in Chapter 9.

Create Walkable Neighborhoods

Smart Growth and Prefurbia neighborhoods design pedestrian systems in very different ways.

Most Smart Growth neighborhood locate walks parallel to the street curb - for walking connectivity, and most place homes front closer to the street edge (curb) than typical suburban development - much closer. Thus, pedestrians walk very close to street traffic. The street 'system' is the walk 'system'. Streets, thus walks, lead to the open spaces and commercial areas. The New Urbanist planner place open spaces (parks) within a 5-10 minute walk from every home. Right-of-ways of streets are typically between 40- and 60-feet wide. There is not much room between the street curb and the walk, the location where trees are often placed - with little room for roots to grow, thus eventually destroying curbs and walks. This creates a major headache to the public works departments who must repair curbs and walks displaced by the roots of the trees as they mature.

Sidewalks close to the curb also have another major problem – parked cars. Residents will always park their best cars in the garage or covered parking, while the less valuable vehicles are likely to get parked along the street, curb side. Strolling along a view of older vehicles detracts from the neighborhood ambiance. Urban blight often begins with decaying car cluttered streets.

Due to emphasis on reduced vehicle usage, 'Smart Growth' proposals include renderings of beautiful streetscapes with few, if any parked cars, which will look worse when the streets get obstructed by parked cars, especially after work hours when residents are home. Parked vehicles also tend to create safety issues because they create a visual block to the pedestrians as they stroll along and cross the street. This is an issue with all residential development, even those in Prefurbia. However, in Prefurbia, there is a much greater opportunity for 'off-street' parking.

Prefurbia separates the walks from traffic lanes as much as possible, thus creating safer and more serene settings that invite a stroll. During LandMentor trainings, planners are taught to design the pedestrian system first, even before streets, lots, or homes are set in place, guaranteeing connectivity.

As explained in more detail later, Prefurbia utilizes meandering walks that are set in public easements when they expand beyond the right-of-way, providing ample space for street trees, without damage the public works department would otherwise contend with. The Prefurbia streetscape takes on a park-like setting. Prefurbia walks widths typically exceed the regulatory minimums (which are usually too narrow for a couple to walk comfortably side by side).

When narrow sidewalks are close to the street, many pedestrians will choose to use the street.

Prefurbia home and lot orientations can easily accomodate multiple vehicular storage when needed. Vehicles are placed out of sight, creating a less cluttered look, and allowing for narrower streets without sacrificing safety compared to 'Smart Growth' where cars are typically parked both sides of a street, and exposed to the weather.

In Prefurbia walks are a 'separate system'. In 'Smart Growth' the street and walking system is combined, as is the case with most suburban planning.

Distinctive, attractive communities foster a strong sense of place

Smart Growth and Prefurbia promote designs that create a sense of place. Distinctive, interesting places are essential to motivate residents to get off the couch and into the neighborhood.

Smart Growth developments require a high level of architectural and landscaping elements in order to create an attractive neighborhood - along a grid. This is because all homes are at the same setback providing no opportunity to create uniqueness. To create a 'sense of place', Smart Growth relies upon a much higher level of architectural detail, not planning, to create unique areas. This higher level of architecture and increased attention to landscape architecture is not cheap, thus the reason there are few (if any) affordable 'Smart Growth' neighborhoods. This is also the reason both Architects and Landscape Architects promote Smart Growth aggressively as they make more money. Architectural detail is also the reason Smart Growth neighborhoods are so inviting, however there is nothing to prevent the same level of detail in a Prefurbia (or for that matter any) neighborhood. The largest difference is cost: a 'Smart Growth' design increases infrastructure compared to similar density suburban design, while properly planned Prefurbia design greatly decreases infrastructure. Thus, more funds are available to both architecture and landscape architecture without a cooresponding increase in price to the consumer!

Prefurbia also demands a consistent standard of architecture and landscaping to foster a sense of place and community. Prefurbia's economic (and environmental) advantages would be impossible for grid-based 'Smart Growth' patterns to achieve. This is because Prefurbia methods reduce street length (and utility routing) by (typically) 25 percent, compared to conventional suburban curved designs, and upwards of 50% compared to 'Smart Growth' design of similar densities. Prefurbia

Usable open space in a suburb.

creates additional open space to handle stormwater surface flow instead of relying on expensive storm pipes, inlets, and manholes. This cost savings would release even more funds for character-building aspects of the neighborhood - and to build more energy efficient structures.

Preserve Open Space, farmland, and critical environmental areas

Smart Growth development in suburbia might result in some open space nearby being preserved, but only if the developer purchased surrounding natural area and dedicated it as a permanent preserve as part of the development. An environmentally sensitive natural area would rarely be used for development. If a suburban or outlying area is 'preserved' via a conservation initiative, it is often part of some large tract the developer purchased at a nominal price because the area was not likely suitable for development, farming, or much of anything else. The point is, just because someone builds a high density 500 unit building as part of a development, does not alone, ensure the preservation of open spaces or protecting the environment.

Often when a developer touts large open spaces it's because they could not develop it in the first place. For example, Hennipen Village in Eden Prairie had large areas of land within the restricted 'landing zone' of nearby Flying Cloud Airport. The marketing material made it appear as if the developer dedicated the undevelopable land as open space. Had the landing zone not existed, there would surely be more paving and rooftop, not park land.

Prefurbia's designs can increase density without reducing space - justifying increased density; thus, if the municipality allows a density increase, they are essentially reducing sprawl. Most Prefurbia neighborhoods create parks and open spaces, and in some cases, permanent preserves.

Strengthen existing communities during transitions

While this book concentrates on suburban settings, the methods can be applied to urban settings or transitional areas of development and redevelopment as explained in later chapters.

Open space made of berm, renders the space unusable for residents.

Using Prefurbia methods, a design team can create an urban neighborhood, at urban densities, with a suburban sense of space, thus increasing their market potential and possibly influencing more suburbanites to return closer to urban centers.

Provide a wide variety of transportation choices

There is no difference between Smart Growth design and Prefurbia as far as encouraging multi-modal forms of transportation choices. In both, it is the combined effort of both municipality and developer (but mostly the government) to come up with solutions to serve the new neighborhood.

Many TND planners tend to promote walking and public transportation by inconveniencing drivers. It often results in permanently destroying the overall functionality of the community. For example, in Minneapolis, when light-rail was built, the planners thought it was a good idea to limit the number of parking spaces at the train stations to force people out of their cars. Bear in mind that in Minneapolis gets 20 below zero for months at a time, so walking, while an option, would be unpleasant. The result was hundreds of cars parked in front of very angry adjacent single-family homeowners near the train stops. This design strategy also assumes everyone is healthy and young enough to walk longer distances, often on icy or wet surfaces.

Conventional subdivision with wide streets in Omaha, Neb.

Transportation planning errors are usually very big ones. They are either very costly to correct or can destroy the livability of a city. Prefurbia strategies work within any traffic systems that are in place or proposed.

Prefurbia directives do nothing to inhibit the use of the car; instead they make it easier and safer to utilize transportation options. Buses or rail is no problem, but those systems are beyond the scope of a single developer, unless that developer is building something on the scale of a city.

This unnecessarily wide suburban street in Las Vegas, Nevada is an example of why new strategies are needed.

As far as new cities (or extremely large developments) being designed by our firm, we are always open to integrating multi-modal forms of transportation that make sense for the climate and needs of the entire population - and which can conform to sustainable policies. Most large scale Prefurbia developments offer a 'plug and play' approach to Personal Rapid Transit (PRT) to provide additional options when PRT becomes more mainstream. This advanced transportation system promised to be a viable solution, but has had a rough start, taking over a quarter century before it's first large scale installations to be constructed in South Korea, Israel, and India.

Make development decisions predictable, fair and cost effective

Again, New Urbanism requires a high level of architectural and landscape elements to succeed. Because of the reliance on strictly defined patterns and dimensional controls, when some elements are deleted or not delivered as promised, the house of cards falls down, as seen in Figure 4.1.

Elements built incorrectly in Prefurbia also have negative results, but Prefurbia does not require the same strict architectural standards, thus it is rare for the design to fail on architecture alone.

Figure 4.1 is an example of an affordable Smart Growth development. Take out the front porch to save money, and a near-future redevelopment situation will occur.

Figure 4.2 is an example of Smart Growth where every unit is placed precisely the same. How is this style of planning any more or less monotonous than the typical suburban subdivision?

Figure 4.3 depicts a New Town Center in a Smart Growth development. Porches (stoops) hover above the street, lacking "human scale". How "smart" is this 'growth'?

A strict set of rules inhibits innovation. President Obama claimed innovation as a key component of the U.S. economic recovery. *"Future economic prosperity depends on building a new, stronger foundation and recapturing the spirit of innovation. Innovation has been essential to our prosperity in the past, and it will be essential to our prosperity in the future"* Prefurbia is innovative and

Figure 4.1 The results of Smart Growth with a few key design elements missing during implementation.

Figure 4.2: White Gables — a New Urban Development in South Carolina.

its potential is yet untapped. It does not follow a finite set of 'dimensional' rules. Its design principles foster innovation, not restrict it.

Encourage community and stakeholder collaboration

Throughout this book we bring this issue up. In our own design practice, we place the homeowners and local business owners' needs above all. Only then can we serve both developer and municipality best. The residents' and business owners' stability is the municipalitys' stability. The citizens' desire to invest in the homes and businesses that the developer created, over other options (i.e. moving out of the city), assures the financial success of the developer and the economic development interests of the municipality. As we painfully felt in the housing crash, when housing fails, it hurts everyone.

Avoiding the term "Green"

We intentionally avoid using the term "green" throughout this book. We do this because green is such a generic term (recycling trash as an example), or a service (such as eco-friendly child care), or a product (geothermal). Land development is simply one of a thousand platforms to be 'green.'

But mostly, we avoid the word green because not everything represented as 'green' delivers on its promise. This 'green-wash' was one of the main reasons the previous "Carter Administration Green Era" failed. Therefore, we use the term "sustainable," thus inferring that it is also green.

Every new development gives us an opportunity to exhibit sound environmental stewardship. Every land development will affect the environment. Conventional subdivisions, New Urbanism, and Prefurbia will all have a negative impact on the environment compared to undeveloped ground. The question is, how much 'negative' impact will each of these design options have?

Green by definition is vague, and certainly in the eye of the beholder. If you fertilize your 'sod' lawn with a product that pollutes less, it could be considered green, but someone who uses low-impact landscaping requiring no fertilizer, might not consider any form of sod as being green.

Green can also be unsustainable. For example, a civil engineer on one of our developments decided that we could use the area below 26-foot-wide private drives as drainage basins instead of using natural surface detention ponds. To cut out the earth five-feet deep (below all of the private drives) and then fill it in with rock as a base for the streets and conduit for drainage, was absurdly expensive - adding an estimated $70,000 per home! The project was no longer economically viable, thus, this 'green' solution was unsustainable. The city refused to approve this solution and the development became economically viable again.

Personal green experiences

Back in the first 'green era' during the Carter presidency, I built a 1980's state-of-the-art home (Fig. 4.4): Passive solar, earth-bermed, with a 10kW Bergey wind generator!

With 'passive' solar, the sun heats up a dark brick floor in the home, which in turn heats the home on sunny frigid winter days. The bricks were built upon a thick concrete base which stored heat overnight. This is known as the 'battery'. No complex systems were needed as the home itself became the solar collector. It proved to work well. The City of Maple Grove, Minnesota, where the home was located, had recently passed a wind generator ordinance (1983) allowing a 100-foot tall wind system to be built on a small city lot with just a permit! Likely the first city with such a ruling.

Figure 4.3 Porches hover above a street (lacking walks) in this development marketed as Smart Growth

So, in 1983, we constructed a 100' tall tower with a 10kW Bergey Wind System having 23-foot diameter blades. A quarter century before today's current green movement, we had built a 'Net-Zero' home (it produced more energy than it used).

The neighbors however, were not so excited and waged a war against the city resulting in Maple Grove being the nation's first city to repeal a wind generator ordinance!

Our region (Minneapolis-St. Paul) is known for being both liberal and environmental. Ironically, not a single environmental group or foundation came out to a public meeting to defend the wind generator ordinance. If they had, perhaps the nation's energy policy might look different today.

Figure 4.4 Passive solar Net-Zero home built in 1983.

Years after it was constructed, the city made an enticing offer and purchased the generator from me.

In 1983, the home cost about $121,000 to build. Twelve years later, it was appraised at $186,000. It is an earth-berm home, which in 1983 was promoted as the future, yet just a few years later it became an architectural oddity, reducing it's resale value. As the housing market recovered in the 1990's the neighborhood's overall home appreciation rate (had it been a conventionally built home) should have resulted in the 4,000-square-foot lake front home being worth a minimum $350,000. I had lost nearly $200,000 by going green, *a hard lesson well learned*.

In late 2008, I chose to build green again. This time, as a mandate because of a land purchase from the City of St. Louis Park, Minnesota requiring MNGreenStar certification, a derivitive of the USGBC's LEED program (a version modified for severe cold climates).

The similarity of my two home-building situations is that the housing market downturn coincided with an increase in energy awareness. There were no new green solutions (since I built my last green home 25 years prior), that we found, which offered both reduced construction costs and energy consumption. It seemed that the higher an Energy rating was on an item, the more expensive it became. The choice today still remains, to pay more now for reduced costs later.

With most green ratings, there is a list of requirements the builder must contend with to earn 'points.' There are many 'social engineering' items to comply with. For example, I earned "points" because there was a nearby bus stop and was within a prescribed walking distance to a coffee shop.

A Green Prefurbia Model

The design of the new home demonstrates an expanded feeling of space. The home serves as a model proving that it is possible to provide a large feel in a smaller, more efficient space. Ironically, being near a bus route earned us points, but there is no provision to earn credit for our space efficient design which reduced the building footprint, reducing environmental impact.

Our 3,600 square foot home with a four-car stacked (upper and lower) garage consumed 10 percent less land area (reducing run-off) than our previous home which was a 1936 Cape Cod with a two car garage (2,200-sf). No points for that but points for being able to walk to a coffee shop!

My green certification comes with a Home Energy Rating System (HERS) of 59. This means it is 41% more efficient than a home built to national building code.

Our experience the first winter in our new green home resulted in a monthly utility bill being a small fraction (about 2/3rds less) of what it was in our previously updated 1936 home down the street, yet the livable square footage had increased over 50 percent.

This all relates to the beginning of this chapter – item two of the Smart Growth principles: creating compact housing. Housing needs to be well designed - not just compact.

The difference between the 1936 home and the new one is efficient design techniques. We did not do anything excessive or unusual to obtain the low utility bills. Instead of expensive geothermal systems that receives frequent media attention, we simply paid a few thousand dollars extra for a highly efficient (96%) heating and cooling system with a three phase air exchanger.

In the 26 years that passed between building our first 'passive solar' designed home and the current 'passive solar' designed home, an interesting thing happened. The laws governing windows (demanding Low - e) prevented the sun's uv rays from heating the home! While the window manufacturer did reinstall modified windows in an effort to let more of the sun's energy in, there is still not enough solar gain to provide extra heat. Ironically, the Minnesota State Laws that are designed to prevent energy loss and the sun's rays from destroying furniture, also prevents the sun from providing passive solar heat! Today, there may be more 'glass' options to achieve solar gain.

In addition to the 96% HVAC, a few thousand dollars extra provided a one-inch structural foam seal to supplement the conventional six-inch wide fiberglass batting inside the home. The standard window my builder (CreekHill Custom Homes) used was Anderson 400-series insulated glass, it did not cost any extra, but we did install Hunter Douglas thermal shades which added another few thousand dollars, but offers a short payback period in energy reduction.

As of the writing of this edition, we are entering the 10th year of our 'no-mow' lawn. In general we, and more important, the neighbors were very pleased - initially. However, when a large street tree fell over the yard during a storm and heavy equipment to remove the tree drove over the area, our pristine yard was destroyed. Currently we are exploring options which may be to start over with a no mow or simply sod the yard as the no-mow is expensive to maintain in a pristine condition.

Prefurbia addresses 'green' by providing enhanced efficiency to develop land at a lower cost, releasing funds that can be applied to more efficient systems during home construction.

CHAPTER FIVE
Sustainable Development: A Practical Approach

"There are no such things as limits to growth, because there are no limits to the human capacity for intelligence, imagination, and wonder"

— President Ronald Reagan

Previous chapters concentrated on what has gone wrong with suburbia, Smart Growth issues, the planning and consulting professions, as well as the land development industry. From this point on, this book focus is on innovative solutions that have proven market success. *Solutions that create a more preferred living standard - Prefurbia.*

We use the term 'sustainable' in the title of this book, yet that can mean many different things to many different people or organizations. According to a UN's report, 'It is generally accepted that sustainable development calls for a convergence between social equity, environmental protection, and economic development (the Three Pillars- People, Planet, Profit), yet the concept remains elusive and implementation has proven difficult', because sustainable development has often been reduced to only environmental issues. Without innovative solutions, environmental goals often conflict with the developers ultimate goal for profitability.

Defining sustainability

The **Environment** is the first thing that comes to mind when most people think of 'sustainability'. It is the most commonly intermixed with 'global warming'. The 'greening' of our society through innovative products is constantly debated by politicians and the press who has brought environmental concerns to the forefront. The land development industry is primarily focused on reducing storm water run-off, improving water quality, and increasing energy efficiency in building construction. Prefurbia environmental advantages include both auto and pedestrian centric connectivity while reducing energy and time in transit (*flow*), and less maintenance using natural landscaping solutions; reducing site grading, etc.

The 'profit' term is the bottom line shared by all commerce. We address more than just the goals of a developer. For the purposes of Prefurbia we prefer to use the term **Economics** because there is more to the financial aspects of sustainability than for the developments initial profit. *One can develop a perfectly profitable slum.* If the developer (or municipality) profits but it costs excessively in perpetually to maintain, then the development is unsustainable. If the development was both profitable and impacted the city minimally on maintenance costs, but sacrificed vehicular energy while traveling in the development, or placed a larger maintenance burden on the residents, then it too would be unsustainable. If overall property values could not keep up with inflation, that too would be unsustainable. A good neighborhood design represents a significant economic benefit in a wide variety of circumstances.

We use the term '**Existence**' to replace the term 'People' in Prefurbia terminology, because all neighborhoods should create a sense of pride and accomplishment at any income level, not just for gentrified development. A front porch and tree-lined streets are simply not enough. Prefurbia enhances human dignity, no matter what income strata the resident may fall in. Prefurbia reinvents - not only the suburbs, but also urban development. Cities, environmental activists, planners, architects, etc., do not build communities – land developers and home builders do. Sustainable development cannot happen by itself, it takes a collaborative effort for all stakeholders in the land development process.

Achieving Sustainability

There are many *layers* to design a sustainable neighborhood that fall within the responsibility of those hired by the developer to plan the site.

It should by now become clear why the land planner needs to do more than just layout the streets and then some lots. Without an understanding of all the aspects of the development and construction process, how can the land planner possibly achieve sustainability if they do not possess an awareness of the problems?

Economics:
Must be Profitable
Thriving businesses
Low construction costs
Low Energy Use
Enhanced Values
Low maintenance
Character Building
Hide the unsightly

Existence:
Livability
Pride in Self
Individualism
Space & Views
Community
Health
Security
Investment

Environment:
Preserve natural terrain & vegetation - Reduce impact of storm systems - surface flow
Reduce paved surfaces - Reduce erosion from surface flow - Reduce particulates - Promote natural low maintenance solutions - Reduce Energy Use - Expose Nature To Residents

The land planner who concentrates only on minimums to achieve the most possible density cannot achieve sustainable results.

Sustainable layers

A development cannot be sustainable if it is not approved by the municipality. And following approval, a project would not be successful, nor sustainable, if no one wants to live there, or locate their business to the new project. By including the following 'sustainable layers' into a development design, expedited approvals, increased profits, and environmental responsibility are more likely.

Safety layer

A primary human need is to feel safe. While the land planner may have limited ability on safety as it relates to crime, he/she can have a significant improvement in other areas by design.

There are various graphs, surveys, and research that support a variety of studies regarding vehicular and pedestrian safety. Unfortunately much of the research conflicts with planning-related issues. It is not difficult to find some expert offering an opposite opinion on something as routine as street and sidewalk patterns - *depending who financed the study.*

According to www.globalroadsafety.org the loss of wages, property damage and other factors from traffic related accidents in 1994 represented 4.6 percent of the Gross National Product (GNP) of the United States - today it's much worse. That's right - five cents out of every US dollar went to accidents that could have been avoided or reduced in severity had land planners designed the 'systems' better. According to the World Heath Organization (WHO) book, 'World Report on Road Traffic Injury Prevention', an estimated 1.2 million people are killed in roadside crashes each year worldwide, as many as 50 million are injured. Of these, one thing stands out. In areas where pedestrians and bicyclists intermingle with cars and trucks, accidents are higher. The answer for reducing pedestrian/auto related injuries is clear: separate the vehicular and pedestrian systems as much as is possible. In complex traffic situations which include 4-way intersections and traffic circles, drivers are looking to avoid other vehicles and may be less aware of the pedestrians crossing the same intersections. Prefurbia design methods reduce these multi-modal conflicts.

Environmental and economic concerns should never take priority to safety in design. In particular, reducing potential impact points and high speeds that cause serious injury and death must be stressed. It's impossible to design a completely safe neighborhood that serves both cars and pedestrians. Thus, the land planner can only reduce the number of dangerous situations within a development - not eliminate all of them.

Environmental layer

No doubt developers in the past bulldozed their way to profits, clear cutting the natural terrain. This had a terrible impact on both the environment and developer reputations. Regulations, including the Clean Water Act, reduced some environmental damage, but not enough. There are proven design methods that enhance home values through environmentally responsible design while decreasing development costs.

Smart Growth strategy for reducing environmental impact includes leaving large open natural areas while preferring to compress homes in tighter spaces, to mitigate the effects of growth. While a noble goal, the resulting dense developments have very little organic surface area to absorb rainfall.

Even if higher density were placed in suburbia to preserve open space, it will at best only soften the environmental impacts of growth. The reason is that land set aside, unless dedicated as permanent open space, or preserve, will eventually be filled up with more rooftops and paving.

The residents of individual clusters of development will still need to get to and from work, services, schools, and conveniences. In other words, 2,000 suburban residents will wait in line on the highway to travel 30 miles to work regardless if they live on 10,000 square foot lots or 4,000 square foot lots on the same 600 acres, even if 300 acres of that land is dedicated for conservation.

The environmental layer must address low impact storm water management. Later we will discuss specific methods to incorporate (or avoid) solutions such as pervious pavements and bioswales, that can meet environmental concerns, but be unprofitable. It is possible to reduce fuel consumption through smarter design of local streets to improve traffic flow, as it is possible to design neighborhoods that will encourage walking or biking over driving. Designing with a balance is the key.

Economic layer

A builder who provides higher value will sell more homes than the competition down the street trying to sell a similar sized product with less *perceived* value. A home buyer who cares about safety and the environment is willing to pay a little extra for it. That same home buyer will pay even more if their home is in a neighborhood that has views and direct access to open space - *premiums*.

This picture of suburban Albuquerque is very typical of the development pattern in the southern sections of the USA. Compact lots appear to maximize density and adhere to ordinance minimums, while increasing environmental impact.

When affordability is a primary concern, it is important to consider how price will influence a home buyer to make more responsible choices and, more importantly, how that decision could determine whether the buyer will prefer your neighborhood over another down the street.

Aesthetic layer

Middleton Hills, a well known New Urban Village in Middleton, Wisconsin. Note the intensity (compression) of space and clustering to leave natural areas. Those living along the natural areas (the premium lots) benefit, but those internal to the development lack a sense of space. The intensity of the developed area leaves little pervious surface area.

While beauty may be in the eye of the beholder, a developer and home builder will want to cater to as many potential buyers as possible.

Does a family actually need the US average home of 2,500 square feet? Do they really need to move 10 more miles away from work? Is the emotional need to move up in status more important than the pain of debt, increased commute time and fuel costs, etc? Historically for the most part, the answers reflect the fact that home sales are often based upon emotions. The exception being in the decade before the recession, where people bought homes because they increased in value beyond inflation, with easy to financing, and that home equity was used as a source of additional income. People bought homes not because they loved the house but because it was an additional net worth generator.

This courtyard of a successful starter home TND delivers homes with architectural elements and character not often associated with affordability. The landscaped courtyard (similar to the 'Bays' of BayHome development in Chapter 10) is a refreshing departure from the typical TND.

After the recession began in December 2007, home sales stalled. In places like Minneapolis, Las Vegas, Phoenix, etc. where home prices rose faster than inflation, the values plummeted. But even with compounding foreclosure problems, the over-inflated home values simply went down to the NAHB national average of $264,000 (2007) in most cases. In other words, that 2,500 square foot home in Minneapolis that sold for $400,000 deflated to the national average home price - *where it should have been all along*. The real estate crash in Dubai (see picture below) was far more reaching, destroying the economics of growth in the entire Middle East.

Lessons should have been learned from the housing crash. However, we have seen no evidence of design improvements since the markets began to recover in 2012. What will it take for developers, builders, and cities to change? Without change there can be no progress - without progress we cannot achieve sustainability. If today's consumer is shown an energy efficient 2,000 square foot home that felt much bigger than the 2,500 square foot homes they saw in the past -*they would buy it*. If the neighborhood *felt* more spacious yet was *higher density* than the one they currently live in - *they will move*.

If consumers were provided an inviting, safe and convenient walk system with actual destinations to walk to - *that would encourage a sale*.

2012: Mostly vacant buildings west of Dubai.

If they felt moving meant less depreciation over time compared to the subdivision they currently live within - *they would purchase*.

It is critical that developers and builders deliver superior neighborhoods that competes not only on price, but the dreams of the consumer.

This is a cul-de-sac island in Roseheart (San Antonio, Texas) at dawn when residents along the cul-de-sac wake up to leave for work. It certainly provides a pleasant way to begin the day.

Aesthetic layer
Successful development includes two other elements essential to neighborhood character: architecture and landscaping. For long-term sustainability, architectural elements should lean towards timeless qualities. Avoid trendy designs - the Mansard roof desirable in the 1960's seem downright ugly today.

Architecture and landscaping provide that critical 'first impression'. This is why many national developers spend enormous sums on entrance gates. Most of the time when people comment that it is 'a well-planned development', it has nothing to do with 'planning' - it usually has everything to do with architecture and landscaping.

Conversely, a well *planned* neighborhood that lacks architecture and landscaping is often thought of as a poorly planned place.

Realistic approach
To help readers understand the innovative proven solutions within this book, we touch on many aspects that make up sustainable neighborhoods. This will include some basic knowledge of engineering and surveying discussed in simplistic terms which are easy to learn and understand. A more in-depth foundation for planning, architecture, engineering and surveying is provided within the LandMentor 'system', its trainings, and mentoring.

Instead of broad-brush planning where general street patterns and land uses are designated, we prefer taking a more intimate approach – getting into specific highly detailed planning immediately - at begining design stages of design to eliminate situations that can compromise a neighborhood. With technological breakthroughs in site design it takes very little extra effort to create highly detailed rendered plans and interactive 3D presentations. By supporting VR headsets, we can place anyone inside the development - years before it's completed!

SECTION THREE

Prefurbia Design Strategies

- Land Use and Environmental Conditions
- Transportation Systems
- Coving
- Mixed-Use and Multi-Family Housing

CHAPTER SIX
Land Use and Environmental Considerations

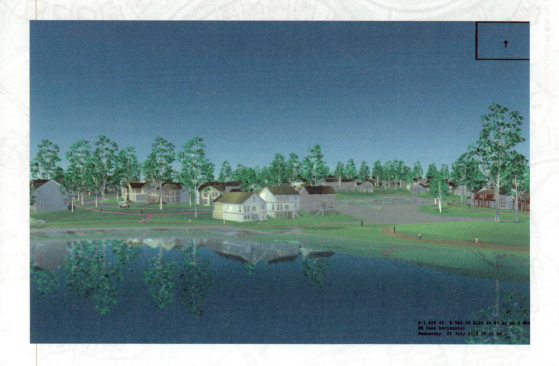

"Suburbia is where the developer bulldozes out the trees, then names the streets after them."

— Bill Vaughan

Critical information such as site boundary with topography (contours), along with water, sewer, and transportation systems, all fall into the required elements of neighborhood design.

These elements have a direct impact on economics, environment, and the people who choose to invest in the neighborhood.

This chapter will explain many of the surveying and civil engineering criteria important to understand for development decision making. This book is not meant to be the end all, but to introduce the information needed to create a sustainable development.

The base material: land

This fictional tale is based on experiences we encountered in the past.

Ralph Dogood, the town doctor, decided to develop land. After looking at quite a few sites that seemed too expensive, he hears of 40-acres marketed for only $30,000 an acre, a price at least $10,000 less an acre than other sites he had been researching.

He knew others were looking at the same property so he offered $22,000 an acre in cash and could not believe it when the land owner took his offer! After all, 40 acres is, well, 40 acres - isn't it?

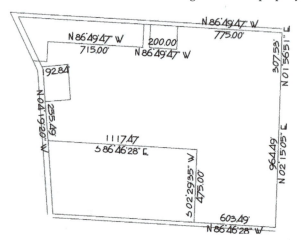

Figure 6.1

From court records he obtained a copy of basic site information (Figure 6.1). Not knowing much about surveying, it looked good to Ralph.

Unbeknownst to Ralph, it is typical that city derived boundary information is not accurate. Mapping may be traced by hand by someone in the city or by an outside vendor – typically the lowest bidder. The area was based upon old tax information and never verified by a land surveyor.

The site includes an odd, yet unusable shape at the northwest portion and an unusable narrow access to a street reducing Ralph's utilization of the land by a third of an acre – still, not a big deal.

The city also had some rough topographic information that Ralph was able to obtain (Figure 6.2).

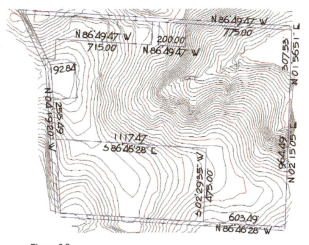

Figure 6.2

It all looked good as he figured it's better to have some hills than a boring flat site. He particularly loved this site because it was heavily wooded and not the featureless flat farm fields he was originally considering.

After Ralph signed a 'quit claim' deed (no guarantee) for the property, he asked the local surveyor to sketch some layouts. The land surveyor investigated the legal aspects of a site, and discovered problems.

Although the city map did not indicate any easements, it was discovered that the local utility company reserved the rights for a gas line 20 years ago when they did their long range planning. They had not yet reached the site, so nothing 'physical' on the ground indicated such an easement, but it was there nonetheless.

This easement bisected the site, as shown in red on Figure 6.3. Sometimes easements may not be seen as limitations, however, they can have a dramatic effect on the layout and development of the land. A utility company defining an easement does not typically consider the impact on the future land development when they construct their underground pipes or overhead powerlines.

Still Ralph was not upset – he paid almost half of what the other land in the area cost. The surveyor said that they needed to do a wetland survey. Ralph assured them that the site was not wet other than a small spot at the northeast corner. Ralph did not understand that a "wetland" is based upon soil types and plant materials, and can appear quite dry. To Ralph's shock the surveyed wetlands wiped out a third of the site.

The county regulations required a 40-foot buffer around wetlands, which took out another three acres of useable land (indicated in yellow on Figure 6.4). Finally, the engineer's estimate of $400,000 to grade the site for development because of the steep slopes, (which would also wipe out every tree in the buildable area) destroyed any chance of developing within a reasonable budget.

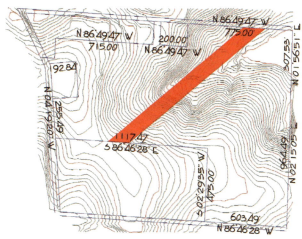

Figure 6.3

Looking at the drawing of the easement and wetland, it would be extremely difficult to make any economically feasible layout of the land. Ralph was lucky that his surveyor had insisted that the site be investigated before any concept plans be done, and an engineer gave an honest opinion as to the cost of grading. All too often a consultant will continue billing, suggesting to the developer that perhaps all is not too bad, since preliminary site analysis work can generate significant billable hours. Luckily Ralph was able to sell the site for $20,000 an acre, reducing his losses.

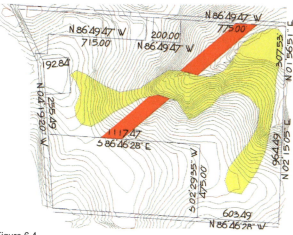

Figure 6.4

While there are many successful development stories, there are also many developers who have gone broke knowing about problems at the beginning, but who forged ahead away. In aviation there are many stories of pilots who made a collection of small mistakes resulting in the plane spiraling to the ground. similar to land developers small but bad decisions leading to unprofitable development or bankruptcy.

Figure 6.5 Saddle Creek in Carmel, Indiana by Pulte Homes - one of the first Coved neighborhood designs.

The property
It is also rare that a site has the exact dimensions that will allow lots to be placed at the precise minimum lot width and depth using conventional design methods. The shape of the land determines the efficiency of the site.

A large rectangular site is more ideal for conventional and TND planning. As soon as a site boundary takes on an odd shape, the designs options in Prefurbia become more efficient for both residential and commercial solutions. However, site shape is not the only constraint.

Not too long ago, developers filled in the abundant swamps on their land. Today, these swamps are called 'protected wetlands' and cannot be filled in - legally.

Wetlands should be considered as a physical boundary. No matter how they initially look or are drawn on a city map, the only wetland boundary that should be relied on is one that has been delineated by a professional and then surveyed to locate it exactly on the site. Wetlands and shore lines are unlikely to be the best shape to conform to the rigid dimensions of conventional or TND designs. Shorelines often come with additional restrictions and setbacks.

Until environmental restrictions became ever more restrictive, developers simply built over sensitive areas.

Easements
A pipe line may only be a few inches wide, but the easement required for it can wipe out several acres. While some utility companies are reasonable, others may request you not use the easement areas for any type of development. Thus, the easement acts more of like a right-of-way, reducing

property value. If the utility company has an easement, it is just that, a swath of land to maintain the utility within that easement, not full ownership rights, otherwise the utility company should pay for full price for the land, not just 'easement' rights.

The problem for site design is how to create a plan that hides the easement as an obvious element within the neighborhood. As an example, residents living in Saddle Creek (Figure 6.5) are never aware of the sanitary main easement that runs diagonally through the site.

Wastewater - Sanitary sewer systems

Wastewater travels downhill using gravity. Waste leaves a home in a small pipe and joins larger pipe that also handles the neighbors' waste. Since gravity moves the waste, the pipe needs to be set at an angle steep enough for flow, but not too steep as to create an abrasive action that can prematurely wear out the pipe.

In a perfect world, there just so happens to be an existing city sewer pipe located adjacent to the site at the correct elevation and location to connect to. In northern climates, the pipe needs to be deep enough to be insulated against freezing and accomodate basements.

If basements are the norm, the pipe journey will begin about 10 feet deep and get deeper as pipes continue along their path. It is not uncommon to reach depths of 20 feet or more.

What if that pipe breaks or needs repair? How can a service man reach it? Enter the "manhole." The repair man steps down into a manhole to access the pipe. There is a limit to the distance that cities will allow between manholes because of the need to access the pipes as well as a limit to how deep that manhole can become. Today, it is possible to repair most pipes without digging them up through robotic devices that can crawl through the pipes that remotely seal the leaks and breakage. Most cities do not allow bends in sewer mains (some do), so any change in direction means another very expensive manhole needs to be constructed.

If the pipe that is needed for a new development to connect to is at a higher elevation than the point it needs for gravity to work, then sewage must be lifted to a higher elevation to begin its downhill journey again. This sewage 'elevator' is called a lift station. To handle a large quantity of waste without failure, the lift station isn't a cheap item.

Eventually the waste leads to a treatment system where, in simple terms, microbes break it down. Waste is cleaned enough to enter the natural drainage system. This "traditional" centralized system has existed for many years.

Today, *decentralized* systems are becoming a viable alternative. These compact systems can provide collection and treatment for a group of units, or even for an entire neighborhood. What makes them decentralized is that the waste is not collected and conveyed to one centralized treatment facility for the entire city.

If an engineer does not want to try anything new, they will make up stories (i.e. lie) to sway developers to conventional subdividing. In 2012, we got a call from a Canadian developer saying they needed to straighten out all our streets because the city engineer (an employee of the largest consulting firm in Canada) told them curved sanitary sewer pipe costs $51 a foot! *If only there was such a thing as curved sanitary sewer pipes!* In San Antonio, one of the largest engineering firms lied to our client telling them our design with 27% less street was too expensive to build. When confronted with the truth, the developer ultimately made the right choice and went with our more sustainable design.

Waste treatment
Here are three basic options:

1) Septic field:

Waste flows into a holding tank where the large solids are separated and the sewer water continues into a series of pipes with small holes along the bottom. This allows slow seepage into the ground which was layered with different materials that filter the seepage.

The septic field is designed so that if some holes become plugged up, the sewage is redirected to unplugged areas. In theory the plugged up areas eventually become useable and the field can work indefinitely. However, if the entire field gets plugged up a new field must be built. This is why most towns require large lots for septic fields. When opponents to sprawl see these large rural lots they often assume these people are wasting land when, in fact, they simply are complying with the physical requirements to service septic fields.

2) Conventional treatment plants and de-centralized systems:

When a development is large enough, it becomes feasible to build a facility that serves a large community. This option requires a treatment plant to be built which costs millions of dollars and is typically built by the city. In most instances the management of the facility is turned over to the community. Technological innovations continue to refine the various options. Smaller decentralized systems have an advantage: the developer can achieve higher densities in rural areas. This can mean a reduction in the initial purchase price of the land compared to land with urban services, offering more affordable housing.

3) Low-pressure, vacuum, and pumps:

Gravity flow systems rely on soils that are easily trenched which have slopes that lead downwards to an existing sewer line to tie into. A viable alternative, especially when the connection point is located above the proposed development and the soils are difficult to trench (such as rock), is the low pressure sewer. With this system, a grinder pump at each home or business sends the sewage under pressure in a flexible small diameter pipe that also eliminates the need for manholes. Pressure sewers sound like a perfect solution, but the grinder systems are not cheap, so the overall costs between gravity flow and pressurized systems are not that far apart. Placing two or three homes per grinder is a great way to make the economics of pressurized alternatives far more attractive compared to gravity systems.

Stormwater – at the macro level
Stormwater and natural drainage are well documented concerns in land development.

Have you ever noticed how America's older cities rarely flood? Were the engineers 100 years ago smarter than today who let their 'Storm Master' software design for them? Years ago engineers had a secret. They did not know how to precisely size systems, so they just oversized everything! A refreshing rain from a passing cloud was not the problem. It was the massive storms that create the angst.

With natural ground, untouched by man, a storm drops the exact same amout of water as developed land, however, nature's undeveloped land soaks up much of it. Plants absorb quite a bit, and most soils can handle a decent amount of infiltration. Some of the moisture evaporates back into the air. This natural stormwater management system feeds our ponds, creeks, streams, and rivers that have *naturally* sized themselves to handle the volume that comes their way.

When we develop land, everything changes – drastically. A developer takes some of the ground (say 100 acres in the suburbs) and converts 40 (or more) percent of the land surface to paving and rooftops. Solid surfaces do a poor job of absorbing water. The remaining land cannot absorb the extra 40 percent of water and sends it downstream - faster. This is similar to turning on your faucet just a little more than your sink drain can handle. One development after another opens up that faucet until it can be opened no further. Our streams and rivers were designed by nature to handle only so much water. Animals, (including humans), and plant life depend on these natural systems. This is why on-site storm water management is the norm.

Stormwater – at the micro level

Streets typically also serve as a conduit for rain water. Residential streets trap rain along the curb directing it to inlets which are connected by pipes to manholes (the same expensive type of manhole the sanitary sewer system uses), which contains a larger (extremely expensive) pipe that is connected to similar storm water systems in adjacent developments. Each system contributes more volume, requiring a larger pipe or increased slope to handle the volume. These are big pipes, some so large you can drive a bus through them. Large expensive pipes get rid of the rain quickly to avoid street flooding. The enormous force of this fast running massive volume of water often outfalls into our lazy, slow running streams with awful consequences.

Detention – retention

Today there are many federal, state, and local laws that mandate the management of stormwater. In many instances this requires maintaining the same rate and volume of run-off onto the neighboring drainage systems as if the site had remained in its natural state.

For the developer, these 'detaining' systems reduce useable land and can be expensive to build and maintain. This cost is *always* passed along to the home buyer - you.

Like innovations made in wastewater, many have been, and continue to be, made in the management of stormwater – too many to go into in this book. However, the typical detention or retention pond to regulate stormwater discharge has become 'old-school.' In addition to consuming developable land, it often destroys the aesthetics of a development and typically wastes great opportunities to create a passive amenity for residents to enjoy. Most of the traditional detention ponds reduce useable green space – and they can be seen as an added insurance risk (children drowning in unattended ponds) requiring fencing and constant upkeep.

Figure 6.6 Westridge Hills rendering - a Prefurbia neighborhood with natural surface flow.

Figure 6.7 A manmade prairie

Surface flow

It is best to use surface drainage instead of expensive pipes. This means that the *initial planning stage* must take into consideration the natural drainage patterns of the site to determine where the open spaces will flow to. Westridge Village (Figure 6.6) is a great example.

Most residents will never be aware the site's drainage is handled through the meandering open spaces, absorbing rain-fall as it makes the journey to a central lake. A land planner without basic civil engineering knowledge is of little use to design such a neighborhood.

Prairie instead of lawn

A man made prairie can help absorb rainwater and cost less than laying sod. A properly planned prairie can attract birds and wild life that can transform a subdivision atmosphere to a more rural-like setting like Creekside Village of Sauk Rapids, Minn. (Figure 6.7).

Rain gardens

Rain gardens can filter out small particles of pollutants that are normally picked up by the flowing stormwater and led down stream. For much of the world, this means the ocean.

When there is no filtration, the Sierra Club web page on water quality tells about the effects of these small particles where the Mississippi River flows into the Gulf of Mexico:

"Every summer in the Gulf of Mexico an area becomes void of life due to severely depleted levels of oxygen in the gulf's water, a state known as hypoxia. This condition kills every oxygen-dependent sea creature within its zone. The 'Dead Zone' varies in size from year to year, but generally it has been growing since 1993. In 2005, researchers mapping the Dead Zone found that it covered 4,564 square miles, an area slightly smaller than the state of Connecticut. In some years it has covered up to roughly 7,000 square miles."

You can see what these rain gardens look like at The Fields of St. Croix (Figure 6.8).

Figure 6.8 Rain gardens can destroy curb appeal and hide good architectural detail, and close space.

Properly designed *and maintained*, they can be quite attractive and add both character and value to the neighborhood. Plants in rain gardens are deep rooted. They live, die and then regenerate. The dead roots deteriorate, leaving a sponge-like quality to the soil that filters out pollutants. That's the simple explanation on how it works. Unfortuanately many potential home buyers see these areas as 'weed-gardens' which may scare builders and home buyers away.

Notice in Figure 6.8 that there is no curbing. Rain falls from the street to surface systems which pass through the rain gardens. It is likely that if all land development had mechanisms to

Life Experience

Oops!

As a pilot I did my usual 'fly over' of the many local sites we planned to see how they were progressing. On flying over Summit Hills in Dassel, Minnesota, I saw what appeared to be a pile of dirt about the size of a Wal-Mart Store. Thinking that was odd, I took pictures of it. The very next day I got a call from the developer saying I needed to come out to the site. He needed to show me something. He specifically asked me why the pile of dirt was there. My first thought was, "had the plan I designed not balanced?" I knew of some stories where the engineering firm's software was misread (user error) and no one catches the mistake till it's too late – but in this case I was really worried for my friend, Steve Sletner, the engineer. I immediately got on the phone and asked Steve to get out to the site. Steve arrived with his technician, who had calculated the grading design. They brought copies of emails from the developer asking if the site could be redesigned to clear 250,000 cubic yards from the site as the city needed it for some purpose. The developer was trying to help out the city. The developer forgot about the emails that the city had deteremined that it only needed 12,000 cubic yards of dirt, an expensive 238,000 cubic yard error! An attempt to help out backfired! Lesson; keep track of commitments!

Figure 6.9

Figure 6.10

filter pollutants from the run-off leaving our land developments, the gulf 'Dead Zone' would have never occurred.

The way our past subdivisions were designed, it's impractical to retrofit raingardens, but they can be used on the ones we are planning today.

Moving dirt

Re-sculpting our land (site grading) to conform to development may not seem like such a big deal.

If a developer buys a few hundred acres of flat open farm field and then moves the earth enough to create an interesting site that functions from a drainage standpoint and offers more premium lots that pays for the earth moving, what could be wrong?

In some cases moving dirt is a wise choice, but in others is not - especially if the site is wooded.

Why move dirt at all? It is often done to allow individual home lots to drain properly. In most cases, a site can be designed in a way to provide drainage with a minimum of grading.

"Dirt" is expensive to bring into or remove from, a site – *extremely expensive*. Ideally a development design *balances* the dirt to be moved. That means for every cubic yard of dirt removed or *cut* from one area of the site, the design should be able to use that yard of earth as *fill* to another area on the site.

Much of the land we develop in the suburbs is comprised of farm fields that have few, if any, trees. Often development plans that contain wooded areas appear to save most of the trees from a bird's eye view. However, if the site grading requires most of the ground to be reshaped, those trees will be killed. A half foot change in grade is likely to destroy a tree. Soil type should factor when deciding how much grading is needed. Rocky soil grading is extremely costly.

An example is the site that Roseheart is built upon. It contained rock and was very wooded. Sitterle Homes, the San Antonio developer, desired to save as many trees as possible and eliminate as much of the grading because of rocks. Figure 6.9 is an aerial photograph of Roseheart that shows how construction does not always require clear cutting and grading. However, it is more shocking when you see that just a few minutes down the road (Figure 6.10) how many trees were destroyed on a nearby subdivision built by a large national home builder.

In 2015 that same national builder requested we redesign a 108 lot site in Prior Lake, Minnesota which they initially abandoned because of a $2 million dollar earthwork estimate. We redesigned the development at concept stage to balance the earthwork and nearly doubled the density resulting in Trillium Coves which bagan construction in 2018.

Debunking land-use myths

A decade after suburbia's birth (after World War II) city planners knew that there were some major problems on the horizon for the growth of the nation's cities.

The 1959 Movie, "Community Growth – Crisis and Challenge" by the National Home Builders Association and the Urban Land Institute, discusses the issues of those days – and their *new* solutions. The movie explains many of the problems that we still have today (but without an environmental emphasis) and tells viewers about new ideas, cluster planning and planned unit development, that will help solve these problems.

Cluster planning

Cluster planning is simple. Instead of larger lots spread throughout the site, smaller, more compact lots are designed for homes and the area saved can be used as common open spaces.

Developers often use the clusters to fill the open space with more housing (Figure 6.11), thus no green space for anyone – just larger profits and environmental destruction.

Figure 6.11 Cluster planning in the suburbs - where's the open space being preserved?

The only residents aware of the extra space in their daily lives are those that could afford the premium price to be located adjacent to the open space. Thus, the majority of residents do not gain any benefit.

Rural clusters

As a recognized conservationist and founder of the Natural Lands Trust, Randall Arndt promotes a concept for "Conservation Development" that takes the cluster theory of dedicated open space one step further - to dedicate that space for conservancy. While a noble goal, in reality, the vast majority of these developments are filled with luxury homes, out of financial reach to the mass market.

Figure 6.12: Rural Randall Arndt Style Conservation Planning

Conservation easements are typically purchased by trusts which may be funded indirectly by tax dollars or contributions. Rural clusters can have huge tax break benefits for the developer, or it can create the possibility of selling the land to a land trust for even greater profits. It would be nice if more of these developments catered to the average income group: the mass market.

Figure 6.12 is an example of a well planned and executed development using this type of design: The Fields of St. Croix in Lino Lakes, Minnesota.

All Prefurbia methods can easily conform with the concept of conservation dedication.

Solar orientation site constraints

Homes themselves can be positioned to reduce energy used to heat and cool them. While there has been a small effort to create home designs that can be oriented for solar advantages, for the most part architects only concentrate on north-south home relationships. A multi-directional effort which includes east-west orientations should be made available to yield more home placement flexibility. Today's more efficient materials and methods of construction make solar orientation less critical.

Solar panels on a home roof has many problems. They need to be clean to work properly increasing homeowners burden. The roof needs repair? That's going to be extra costly with panels attached. Street trees? Not with solar homes, as trees block sunshine when they start to mature. A far more viable and practical option would be to reserve a portion of the site as a 'solar farm' (or wind farm) in an area that would otherwise be less desirable for development. This community system would provide tax benefits to the developer and the energy savings would be averaged to each resident. In certian cases, the solar panels could be multi-tasked, for example, along the south side of a highway providing noise reduction and power.

Life Experience

Lessons learned by the recession

Here in Minnesota our housing crash began many months before the national crash. Our market stalled the day that gas price exceeded $3 a gallon. It was not actually the price of gas that triggered the change, but the steady skyrocketing of housing prices, the increasing commute, and homes that were not much better than those built in the 1990's. The homes simply were no longer a 'value' proposition to the family owning gas guzzling SUV's. In addition to the long commute, suburban densities provided less space in an effort to make economics work because of the absurdly increasing land costs. This all combined to the early crash of the Minneapolis - St. Paul housing market

There were some unique aspects of the Twin Cities compared to other major US Cities. The Metropolitan Council in the early 1990's set an urban boundary to curb sprawl. However, cities outside their control welcomed new development creating excessively long driving distances - *their efforts to curb sprawl made it far worse*. Next, we had every one of the top 10 builders competing for land - bidding up prices which in turn increased home prices far above the rate of inflation. When home prices increase faster than the homeowners incomes, there will eventually be a crash... the question was not if, but when?

When housing markets began to slow down in other cities, our planning business significantly increased because developers and builders began to recognize the same old product they were trying to sell was not going to cut it. *We thought our future was set*.

In the months leading to the national recession we had over 100 active large land developments in the planning and approval stages, with an ever increasing demand for the designs of Prefurbia...

... then President Bush announced the "700 billion dollar problem" and bailed out the banks, who immediately shut down every land development we had under contract - in less than a week. It was decision time, do we take our savings and shut down both planning and software business to ride out the recession or use the down time to concentrate on increasing both innovation and technology investment - risking everything. We thought, how long could a recession last? Maybe one year, perhaps two? Banks and investors were not interested in a company serving the land development industry, we would have to survive on our own savings and use equity in our personal property to liquidate everything.

While we still had foreign planning work, it was nowhere near enough to fund software development and pay overhead. The risk ultimately paid off as the housing market started an upward momentum and developers and builders began seeking better ways to improve value. The downtime allowed us to develop new technological breakthroughs, refine methods in both design and presentation while also creating new educational materials for others to learn and benefit from.

This 5th Edition of Prefurbia was updated in 2018 to include new changes and inform you about newer methods and technology that help overcome the roadblocks to progress.

The risk to invest at the beginning of the recession was justified. We hope to help all involved in the land development industry to bring about a new era of innovation and progress as well as foster collaboration between the professionals that design and produce the neighborhoods.

The recession helped expedite the goals to create a more sustainable world.

CHAPTER SEVEN
Transportation Systems

"Everywhere is within walking distance if you have the time."

— Steven Wright

In the typical planning scenario, the transportation layout (street system) is designed first, before anything else. This is not the way to design a sustainable neighborhood. We will touch on that more later. However, because the streets control the plan layout, this is where we will begin.

From an aerial view, street design seems simple, but this is a three dimensional world. Likely the street is also the conduit for stormwater thus, we need to think about designing the street for drainage.

It is possible to make streets too level with not enough grade such as 'flat' areas in Houston. With little slope to allow gravity to do its work, flooding occurs. If a slope is too steep surface drainage may move too fast, bypassing inlets, or causing vehicles to slide down an icy street, through intersections in winter.

Street intersections are best located when there is flatter ground. Standards vary around the country and around the world. Design grades minimum and maximum are stated in local regulations.

Figure 7.1 Excessive paving features increase environmental impacts.

Sanitary and storm sewers (pipe) typically follow the street, thus, it would be *essential* that the person who plans the site consider liquid flows downhill, not uphill. Unfortunately that is rarely the case with many 'non-engineer' land planners.

Traffic pattern

The shortest path is thought to be between two locations is a straight line, but it is not that simple. We do not travel in a single straight direction from our homes, employment, recreation, shopping and educational services.

The reality is that we need to traverse many different directions and distances.

In the past, most cities were designed using a simple grid pattern as residents walked or used horses and carts to get around. The same grid pattern was used to disperse traffic. To work 10 to 30 miles from where you lived was unthinkable. To travel at speeds greater than 10 miles an hour within city streets was unimaginable, thus, in the 1800's when the grid was the norm. Today it is routine to travel at speeds that did not exist just over a century ago. However, the numerous four way intersections, frequency of conflict points, and excessive speeds from straight streets tend to negate most advantages of the grid. A good reference is the book *Gridlock* by Randall O'Toole that details some of these issues.

Where am I?

Typical suburban street patterns either indirectly leads to the end of the development (aka nowhere) or have a haphazard maze-like pattern creating confusion (Figure 7.2).

If it is easy for drivers to get lost, it will be even more difficult to walk, especially when cities require sidewalks to be built on both sides of the street as the only requirement for 'pedestrian connectivity'. This is one of the reasons people drive through the suburbs instead of using the multi-millions of dollars of sidewalks.

Have you ever tried to comfortably walk with a friend or spouse side by side on a narrow four-foot wide sidewalk? Add neighbors walking and biking in the opposite direction, what happens?

Most end up walking and biking in the street instead of using the sidewalks. Or, they simply jump into the car to visit the neighbor just a block away.

Streets leading to nowhere

Enter this development (Figure 7.2) and you will think it's just a long cul-de-sac. Instead, it leads to a dead end. The curved street segments lack 'flow'. There is no connectivity, to traverse from areas on the left to areas on the right. It would require a drive, a long drive. Walks are provided but with no direct cross connectivity to shorten the distance.

Loop-de-loops (streets that loop back into themselves) are to be avoided - seen in (Figure 7.3), it may look cool on the plan, but on the ground it only adds to spatial confusion. We have seen planners think these are a good idea when designing 'coved' plans -they are not! The curves in conventional subdivisions rarely exceed 90 degrees. With coving, however, curves exceeding 180 degrees is routine.

Segmented block sections are to be avoided. A neighborhood should be about connectivity, flow and spatial expansion. Cul-de-sacs do not need to be avoided as long as emergency and pedestrian connectivity is considered.

Figure 7.2 Meandering streets that lead to nowhere in particular.

Figure 7.3 Streets looping into themselves cause confusion on the ground.

There are unique ways cul-de-sacs can be utilized in an organic design, illustrated in a few more pages.

Flow and traffic patterns
All of the previous examples in this chapter lack the continuous vehicular 'flow' of a cohesive "neighborhood". This is best explained as the ability to enter and safely traverse the neighborhood with a minimum number of stops and turns. The example photographs shown on the prior pages, have a significant number of conflict points and very little connectivity. All of the built examples have sidewalks with little actual 'function' except that they adhere to local regulations. Design priority was not given to pedestrian systems, nor are the systems convenient to use. The science of 'flow' is an extensive subject taught within LandMentor and it's related trainings.

None of the previous examples shown have open space visible from the street. All have the same walk and street widths, no matter how much or how little traffic volume the individual street was intended to be used for. This "one size fits all" for streets and walks makes no sense, and is very wasteful, yet city after city that is exactly what is required - worldwide.

Design elements to be avoided
To achieve less environmental impact, there are several design elements that should be avoided.

Brows (the small loops off of a street serving just a few lots), or medians consume huge areas of land yet separate open space from the homes should be avoided. Insignificant landscaped islands are often not useable to residents. These also increase paved (impervious) surfaces and the length of expensive curbing. The continued maintenance expense (especially in snow country) never goes away. If a planner must include brow-like elements in neighborhoods, they should use narrow one-way lanes and be more organically shaped to avoid monotony and maintain flow of traffic - critical to reducing vehicular energy, i.e. fuel consumption.

Sizing the street and walks
One size does NOT fit all situations. In some suburban cites local paved street widths are still 40-feet wide or more, especially in North Dakota where excessive paving adds to their flooding problems! Why? Because the fire and police departments convince the council that any smaller dimension is unsafe. Where do the rules of thumb for these absurdly wide streets come from? Perhaps they are a throwback from the days when homes had no garages and residents typically parked on the streets. Today with three car garages and large driveways this is no longer true. In Wisconsin, during a recent submittal, the city engineer insisted wide streets were safer. So I googled: 'Are wide streets safer,' and the only results were those that said narrow streets are safer! How unfortunate that beyond any logic, the City Council as well as the Planning Commisioners and other city staff feared going against the city engineer who previously was educated and experienced in storm water management - not in street transportation engineering.

Fortunately, most cities have become smarter and have adopted, or are adopting, much narrower widths. It just makes one wonder why the police and firemen in the vast majority of U.S. cities can maneuver easily within 28 feet of street paving and others need much more space? Are their fire trucks 40 percent wider?

Even with sensible widths – why not adopt a ***variable width system?*** For example, Woodbury, Minnesota, now has a minimum 28-foot standard for local streets and a 26-foot width for cul-de-sacs. The width of streets and walks should be determined by the actual volume they will handle. A foot in width may not seem like such a big deal, but in an entire neighborhood it could have a noticeable impact on home pricing, as well as reducing environmental impacts.

Caution should also be taken as to not create a situation where streets are too narrow or make streets too narrow to making it a headache to drive.

CHAPTER EIGHT
Coving

"Everything that can be invented has been invented."

— Charles H. Duell, U.S. Commissioner of Patents, in 1899

    Coving is a design method in Prefurbia that increases space and 'scale' along the streetscape delivering a 'park-like' feel along home fronts. Coving was made possible by utilizing a more efficient street pattern than simplistic conventional suburban design.

    By definition, a cove is an indent in the shoreline of a body of water. In land planning, a cove is defined as the breaking of the parallel relationship of house and curb, replacing the standard home setback pattern with one that sets each home along a fluid curving line, unique from the street pattern. This forms an indent into the building setback – a cove.

    The reduction in street length can be significant, but a typical coved based design should result in approximately 20% to 30% less street length, maintaining density, while adhering to the existing regulatory minimums, as compared to standard suburban subdivision design.

The shaded area illustrates the concept of coving delivering a significant increase of space along the home fronts.

Coving's magic requires a rebalance of space allowing density to be maintained with less infrastructure length. By adhering to and exceeding existing regulatory minimums, the average buildable area per home increases. Coving from an architectural standpoint will be discussed later – here we will concentrate on elements related to traffic. Many of the methods introduced here could be applied to conventional suburban and Smart Growth strategy.

In Chapter 7, some traffic advantages of coving were presented. In theory, coving could be used in any form of street pattern, even within traditionally straight grid designs. However, without combining coving with the proper street pattern it can become inefficient.

What coving provides when properly designed:
- Increased space without impacting density.
- Less space taken up by streets.
- Showcasing of home fronts - greater "curb appeal."
- Eliminates monotony — the neighborhood looks more interesting.
- There are more "premium" lots, i.e. lots with a good view.
- The ability to increase value with architectural blending and shaping.
- The development costs less to construct, compared to conventional subdivisions.

What coving is not:
- It is not Prefurbia - it's only a single geometric model of many pioneering innovations.
- An end all solution for everyone – no single solution is a savior.
- A pattern that can fit all sites (however, we found most sites benefit from it.).
- A guarantee to increased density (although on some occasions, this occurs).

TRANSPORTATION: ROADS AND SIDEWALKS

Prior to 'Coving' with all forms of housing, to achieve density, you need the maximum length of street frontage. This is because the home and street parallel each other. The result: much of the site is consumed by street right of way. In other words to achieve density, it is necessary to fill the site with streets. But building more streets leaves less area for the lots, creating the smallest possible lot areas.

The developer is left with huge construction costs, which are passed onto the home buyer and the city perpetually maintains the *maximum* amount of street. Much of this is because streets must hug as close to the boundary as possible, to achieve the maximum lot density allowed by ordinance.

A 'coved' based street pattern takes a very different approach. Instead of the street hugging the boundary, it does something that seemingly makes no sense – a street pattern that pulls away from the boundary.

Coving uses a winding pattern that reduces the amount of intersections, and can significantly improve flow. The reason density is maintained is the building setback line *length* is stretched because of the difference in street and setback pattern.

The conventional plan shown in Figure 8.2 has 244 lots. The same site re-planned with coving (Figure 8.3) yields 241 lots. However, the coved plan requires 42 percent less street length, (both plans use the same right of way and paving widths). The average lot size on the coved plan is 40 percent greater, which allows many of the lots to have larger homes. In addition, on the coved plan, front yard and rear yard distances between the homes is greater.

Figures 8.2 (above) and 8.3: represent an original plan and re-designed coved plan of Cantura Cove in Mesquite, Texas.

Early coved street design and traffic patterns

The Meadows at Saddle Creek (a Pulte Homes community) in Carmel, Ind. (Figure 8.4) is a typical example of early coved design. Most residents navigate non-stop through the neighborhood to their home. This early coved subdivision design while better than the grid, but is far from perfect.

Four foot sidewalks parallel to both sides of the street and a wide path along the internal parkway at first seem nice, but there are no north, south, east, west shortcuts! Walks and trails abound, but they do not furnish any cross connectivity within the development. Most of the common space is hidden from the street. The common open spaces are rear yard based, with plain home rears and a variety of fencing, some of which will age poorly. The open space is only visible to a few home buyers with the financial strength to afford to live abutting the commons. Engineered detention areas lower rainfall flow rates, but they are conventionally designed, using excess pipes, inlets and expensive materials, and lots of green space - in lawn.

From a 'mathematical' perspective, The Meadows of Saddle Creek is a winner. With larger lots with less street and a neighborhood that feels far more open than subdivisions of similar density.

When Pat Byrne of Pulte Homes decided to use this new 'coving' pattern (in the 1990s), it was a fresh idea that had not yet proven itself. The chance he took paid off by creating a development that provided higher value, more open space with trails, and happy residents. Pat was later instrumental in developing one of the most exclusive coved neighborhoods, Prairie Creek, in Kildeer, Ill.

Figure 8.4: The Meadows of Saddle Creek in Carmel, Indiana.

FLOW: Coved traffic pattern with greater efficiency

Placitas de La Paz in Coachella, California, represents a destination-based traffic pattern. No matter how you enter this neighborhood, you are lead to a highly visible central park area.

Note the walks through the blocks, making it easy for pedestrians to safely and quickly traverse the neighborhood. (Figure 8.5).

At the time we proposed this neighborhood, the required minimum street paving width was 41 feet and the right of way was 60 feet. We were able to get approval for a 50-foot right of way and 28-foot pavement width. In addition, we were permitted to place sewer manholes in the back of the curb instead of street centerlines because of the aesthetic advantages of this coved design. Total reduction of street surface was 40 percent compared to the previous grid proposal!

As you can see in Figure 8.6, all streets 'FLOW' directly from each main entry point to the park. The open space does not serve as a "traffic circle," as by the time one reaches the park, the residents are already at their destination or they live along the park. Note the 'brows' which we now eliminate.

The cul-de-sacs are not closed. They serve as a pedestrian conduit to the park and/or through the neighborhood. The distinct separation of pedestrian and vehicular traffic is a hallmark of modern Prefurbia patterns.

While actual distance from entry point to destination may in some cases be a greater distance compared to a grid pattern, there are fewer intersections (meaning less stop and go). Thus, when designed correctly is more efficient in time and fuel consumption than a conventional subdivision and certainly a TND pattern.

Figure 8.7 shows the actual "before" plan that would have been submitted before the developer Chris Canaday of Canaday Company discovered the coved design option. The result is a neighborhood that delivered *31 more lots*

Figure 8.5: Placitas de La Paz in Coachella, Calif.

Chapter Eight: Coving 91

Figure 8.6 (Top and right): . Placitas de La Paz - Approved Coved Neighborhood and actual aerial photograph.
Figure 8.7: (Bottom and right) Original site plan of Placitas de La Paz and how it would have looked like when built.

with over *40 percent less* street area. This was the first coved neighborhood that held (and exceeded) the "intent" of ordinance minimums, but not the actual written minimums.

The flow of traffic and space

The next evolution of coving design (Figure 8.8) addressed the interaction of traffic concerning pedestrian and emergency access.

Note the oversized cul-de-sacs that are large enough to become a useable detention, landscape, or park area. A standard cul-de-sac with a solid field of concrete or asphalt (typically 80 to 100 feet in diameter) is both unsightly and wasteful. That diameter was based upon a minimum turning raduis of a fire engine. It has no other logic tied to the dimension. Instead of 'minimums' the diameter is increased with a narrower one way lane to serve the homes. This does two important things: it significantly reduces the amount of paving per home served, and allows for an attractive park-like center.

Wider walks through the blocks and at the ends of the cul-de-sacs provide for neighborhood connectivity. These are often designed to provide complete emergency vehicle access, making it possible to serve homes even if a street is blocked off.

Here the 'layers' of design intermix – walks aid the emergency and utility access, open space becomes visible to all enhancing the value. All of this begins to create a 'balance' to deliver a more sustainable neighborhood.

Shape

Organic planning enables the developer to create distinctive neighborhoods with greater aesthetic appeal, while following the natural terrain. Street systems should be more art – less engineering. Creating a neighborhood with character requires more effort than applying standard intersections and cul-de-sacs automated in CAD to attain uniformity.

Figure 8.8: Territorial Trail, Dayton, MN (Pulte) - 'flow' of traffic.

When neighborhoods are designed, effort should be made to create patterns that stand the test of time and give dignity to the residents that will live in the places we design and build.

Walkways

Sidewalks built parallel to the curb line are the norm, even though this pattern increases pedestrian exposure to dangerous vehicular traffic. It is simply another button press in CAD.

Walks as an extension of the curb line make the street seem wider as apparent in this San Antonio, Texas, subdivision (Figure 8.9). It can make the walk unusable when trying to combine curb, driveway aprons, and drainage functions. While marginally walkable, children will choose the street for biking and skateboarding because of the changes in grade to accomodate driveways.

One obvious reason for making the sidewalk adjacent to the curb is to reduce the conflicts with parked cars from extending over the walks. This is shown in the Birmingham, Alabama, subdivision (Figure 8.10) where fronts of garages are too close to the curb line, rendering the walks unusable!

This is a problem when the home front setback is 20 feet (or less) from the right of way line. It is reduced when the garage face is at least 25 feet from the street right of way. Coving, with its spacious meandering flow of openness, reduces this problem. All of the coved neighborhoods we design have meandering walks.

The depth of 'meander' depends upon what the developer and the city feels comfortable with.

Because of the increased setbacks in coved neighborhoods, the rate of meandering can be visually exciting. Meandering walkways make neighborhoods more marketable and attractive. Meandering walks do not add significant cost to develop. The primary change is only in attention to detail. As you can see in Figure 8.11, the meandering walk of Paseo de Estrella in Albuquerque, New Mexico, enhances the sense of space compared to The Meadows of Saddle Creek with standards walks (Figure 8.12). Note that

Figure 8.9: San Antonio, Texas.

there is also plenty of room in Paseo de Estrella to park cars without conflicting with the walks.

Coving requires easements to be indicated on the recorded plat to keep the walks 'public' in areas where the walks expand beyond the right of way. Walk easements are the simple option. Public walks without having to be restricted by the right of way location enhances neighborhood character.

Designed for walking 'connectivity'
Pedestrian systems should lead to destinations and make it as easy as possible to traverse through the site (example Figure 8.13). The meandering walks placed in public access and maintenance easements makes for a complex plat, but a beautiful neighborhood, as seen in Figures 8.11 and 8.13.

Avoid danger: traffic circles and other roadblocks to 'flow' should only be used if absolutely necessary. An entrance circle or landscape element may create focus, but it also becomes a hassle to navigate for everyone, and an increased danger to pedestrian crossing - forever. Design methods that can add neighborhood character without hindering flow is preferable.

Out-of-control patterns
Some curved patterns make for interesting fly-over sights like this one in Chicago (Figure 8.14).

Have you seen the movie "You, Me and Dupree"? Matt Dillon is a land planner working for his new father-in-law, a greedy developer played by Michael Douglas. Dillon dreams of building the ultimate neighborhood.

Instead, Mike Douglas proposes a design made for its highest profit potential – almost exactly like this Chicago subdivision below!

It is easy to curve streets within a development. As sites get larger, residents still need to traverse the region in a convenient manner, so overdoing the curves is counter productive to the efficiency of the design.

Remember something that looks cool on the plan can be awful at eye-level.

Figure 8.10 Cars parked over sidewalks

Figure 8.11 Meandering wide walkway

Figure 8.12 Narrow walk parallel to curb

Figure 8.13 10' Wide Main Trail at Trasona in Viera

Straight street segments (tangents)

Many municipalities require tangents between curves. Until the advent of modern software, plans (recorded plats) had to be checked by hand for mathematical errors – a time consuming effort. Tangents provided easier plat checking the geometry to seek errors - today, there is no logical reason for them because with CAD plat checking manually is not needed.

Figure 8.14: Subdivision within the Chicago area.

Tangents are not necessary within residential neighborhoods that are restricted to lower speeds. Adhering to obsolete tangent requirements will have a negative impact on coved design. With the new coved patterns, we eliminate tangents on every neighborhood we designed since 2001, even though most regulations require them. Luckily, city after city, planning commission and council members as well as city staff recognize boiler plate wording that have no place in todays regulations.

Speed control

Neighborhoods must be designed for safe speeds. Curves are an excellent way to reduce speeds.

Neighborhood speed around 25 miles per hour (mph) plus or minus five mph is the norm. This equates to about a 190-foot radii as a minimum and about 200 to 300 feet or so as more typical for the center of the street radius. Larger radii will also reduce manhole counts and provide safer sight distances than tighter ones. This is a change from early coved designs that used smaller radiuses.

Even with the best performing production cars, it is difficult to achieve unsafe residential speeds with tight curves in a controlled environment – at least without experiencing uncomfortable G-forces. The homes overlayed on the race track (Figure 8.15) show that the curve radii are not all that different from many of the new coved neighborhoods.

Again, balance is achieved by creating an efficient pattern that leads the average driver to a safe speed through the neighborhood.

Curvature can also affect the aesthetics of a neighborhood. Too tight of a rate of curvature may make the homes on the inside of the curve expose too much side surface area. This is not a problem if the homes have architectural detail that wraps the sides, but most don't - even in upscale neighborhoods. Neighborhood design should also hide rear yards and plain house rear walls from passing internal vehicular traffic.

Also, too tight of a curvature can present another hazard, with reduced sight distances.

Life Experience

Aston Martin vs. Curved Streets

Rick Harrison had the experience of taking an Aston Martin performance driving course at the Michigan Proving Grounds with a professional race driver. The first course was to teach vehicle dynamics at very tight turns on a short course.

Rick found it difficult to accelerate beyond 35 mph as the G-forces and rate of turns were beyond his comfort level. Figure 8.15 is a coved plan overlayed on that same track patterned after a track in Lommel, Belgium.

Other courses on the grounds with more generous radii and longer straight sections allowed him to comfortably obtain near 90 mph in the curves and up to 160 in the straights (tangents) between curves.

DESIGNING SPACE

The concept of pull-back to create premium locations

Not all sites are created equal. Some sites are simply suited better for residential development than others. Spectacular sites offer views that will enhance the living standards of the residents that access expansive views. These views could be natural features such as water front, wetlands, lakes, valleys, etc. They could also be man-made features such as golf courses, parks, rain gardens, etc.

The typical method for dealing with a value added feature is to align homes along that feature and charge large premiums, as shown in this Las Vegas, Nevada, golf course development. (Figure 8.16) A very small percentage of the homes actually experience the golf course – the others have no increased value. Few visitors, and most residents, will never be aware this is a golf development!

Overall, the developer left a lot of money on the table with this

Figure 8.15: Combining coving with an image from Google Map™ showing the Ford Proving Grounds in Romeo, Michigan.

design. By pulling homes *away* from the golf course, more people gain direct and indirect views, creating higher values for many more of the homes.

An example of this is shown in Hutchinson, Minnesota. (Figure 8.17) Shown is Phase Two of Fairway Estates. In Phase One (both the phases were similar size) only eight lots were on the golf course, but here we have 20 direct and indirect golf lots. The detention pond handles both first and second phase drainage and yet the density is the same.

Figure 8.16 - Only a small percentage of residents will view the golf course nearby.

Openness for all

The first coved development, Eagle Pass, had setbacks that increased from the standard 30 feet to as much as 60 feet deep – not nearly enough to develop a noticeable sense of space (scale). Be prepared to create depths as great as 1.5 times the lot width to transform cookie cutter plat designs into a park-like streetscape that creates a 'park-like' feel.

Figure 8.17: The coved plan (left) and the aerial of the coved plan at Fairway Estates under development (2006).

Figure 8.18 represents a park in the front of coved homes in Roseheart, a neighborhood built by Sitterle Homes in San Antonio, Texas. The street itself becomes 'the park' and all homes have a significant increase in value. Smart Growth principles would promote a park within a 5 to 10-minute walk from each home. With coving, while a local park would also be a plus, the goal is that all residents live along a park. This encourages residents to spend more time on the social and more public (street) side of the home. Yet, all homes in a coved neighborhood will still offer a more private and secluded back yard.

With Lennar Homes' Settler's Glenn (Figure 8.19), a coved neighborhood in Stillwater, Minnesota, the variation in setback along the main street and a cul-de-sac creates an organic flow of space. Every home is either *on* the increased space or *adjacent* to it.

The minimum lot size by regulation was only 7,000 square feet with 20-foot front and rear yard minimum depths. Needless to say, we achieved a considerable increase in space – all while maintaining density due to the rebalancing unique to coved design.

Figure 8.18

Scale – enhancing space

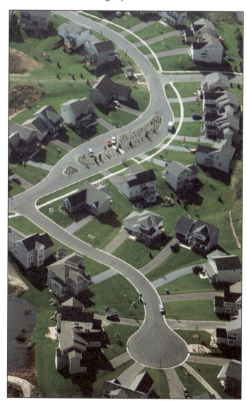

No photo can give the sense of space coving delivers, however, with LandMentor 3D and Microsoft based VR headsets, we can accurately replicate the feel of space of the finished neighborhood at initial stages of design! There is an enormous sense of space when the variation of meandering homes is quite aggressive. Within just five to seven homes, scale is felt from close to far away from the street. This scale change enhances the sense of space.

The opposite is true when the rate of meander used on the setbacks is too gradual. When all homes are deeply setback, there is no scale reference and all that would be accomplished would be long driveways, with no sense of how much space is actually there, resulting in waste. Figure 8.20 is a picture of Pheasant Run in Otsego, Minnesota. Note the homes along the outside of the curved street. The approved preliminary plat had an aggressive rate of meander inside and out. The civil engineer simply decided to make changes destroying the sense of space we designed.

Figure 8.19: Settler's Glenn is an early coving design, which is why the cul-de-sac is solid pavement.

(Remember the beginning of this book about how often engineers take the liberty to change the design without consulting the planner?) There is little sense of the space while driving through this development. Because the homes directly face the street, there is little difference from this coved and 'garage-grove' subdivisions. Note also the lack of any pedestrian system – common on early coved designs.

Figure 8.20: The Pheasant Run development in Otsego, Minnesota. (Not properly Coved)

Angles – expanding views and highlighting architecture

Aggressive rate of setback change, along with homes forming components of curves, equates to homes being on aggressive angles in relation to the curb line. This has the effect of a significant de-emphasizing garage front architecture. In Figure 8.21 of Pulte Homes, Meadowbrook Village in Fishers, Indiana, the far home faces the view from an approaching car rather than facing directly towards the street.

Homes that parallel the curb, and thus, parallel each other, limit space. All views from within and outside the home is the same distance. For example, if the face of the homes are each 110 feet apart and/or the rear of the homes are each 70 feet apart, then that view distance is the limit of space. That limited space is made worse when the neighbors adds a fence or shrub row. The angles of the homes along a coved setback may seem minor, but make a difference in expanding views.

As you can see in Figure 8.22, Wedgewood Coves, a golf community in Albert Lea, Minnesota, there are vast distances of views created for each home within the development by their angle and setback. This translates into a decrease in the feel of density – not just from the outside, but from within every home.

Streetscape sculpting is where the shape of the setback leads the eye to follow along the home fronts toward a view of a neighborhood feature. The added visual impact when a prospective home buyer drives through the neighborhood provides the home builder with a marketing advantage.

Streetscape sculpting leads to views of both the golf course and ponds behind the homes. These would typically be hidden in conventional design.

This form of flowing viewsheds allows the open space to be enjoyed by all residents and visitors.

The increased view vistas is an extension of flow of traffic concepts covered in the last chapter.

In Paseo de Estrella, a DR Horton Neighborhood in Albuquerque, New Mexico (Figure 8.23), note how the space flows between two regions of the site through the park. Residents can easily walk through main sections of the neighborhood. Paseo de Estrella won the New Mexico Residential Neighborhood of the Year award in 2005.

Figure 8.21: Meadowbrook Village.

Figure 8.22

Figure 8.23

Copper Ridge in Billings, Montana (Figure 8.24), developed by the Oakland Companies, clearly shows the concept of using home shapes to create a flow of space separate of what can be created by a strict home/street relationship. Note how all streets and walks lead to the central park. It's no wonder that in 2012 it was reported to be the best selling neighborhood in Montana

Long driveways

When presenting the first coved neighborhoods, carousels filled with slides (this is before laptops and projectors) were brought along to counter arguments expected against coved design. In particular, photographs of older Minneapolis and St. Paul neighborhoods where people drive (along driveways) at least 120 feet along the side of their homes to reach their rear yard garages were on hand. An example is shown in Figure 8.25.

It seemed that no one considered the snow removal maintenance a problem with those older homes, many were in upscale neighborhoods.

Not once was that carousel labeled 'long driveways' ever needed.

The homes with longer driveways are the most premium. Now that there are many coved neighborhoods built, one observation is the street gets parked on much less with coving compared to conventional planning designs. This is because of the longer distances from the coved setbacks. Car clutter that is a blemish of the suburbs is softened (or eliminated) by using coving.

Another requirement of long driveways is that in order to

Figure 8.24

Figure 8.25: St. Louis Park, Minnesota.

minimize impervious areas, they need to taper to narrow widths as shown on this streetscape of Cantura Coves, a DR Horton Neighborhood in Mesquite, Texas (Figure 8.26.) It may be a surprise to learn that this photo represents a neighborhood with a density of 3.6 homes per acre!

Porches and neighborhoods

There is no doubt that porches encourage neighborly interaction While a style or type of architecture is not dictated in coved designs, there is something about a front porch that conveys – 'neighborhood', not just a 'subdivision'. Prefurbia encourages full useable front porches to connect the interior and exterior spaces.

Figure 8.27 is from Lennar Homes' Settler's Glenn. Notice that this suburban home not only has a front porch, it also has three garage spaces! Plus, the home just to the left of it has a large front porch, living space along the street side and, yes, three garage spaces.

Prefurbia demands both architectural and landscape control. There is justification for this as the development cost savings can be applied to these items which add character, as seen in Figure 8.28, a coved neighborhood by Jim Schmitz of Homestead Traditions Development Company, Inc.

Excessive width

Unless the homes are massive , the spacing between homes in some suburban areas is excessive. While this problem has been written about in many other books and articles, one of the advantages of coving is that few home sides are parallel. This means that if homes have their closest possible points 10 feet apart, the average distance will be much greater. With conventional design, 10 feet is exactly 10 feet .

When coving is used, determine the largest possible home pad size, then look at minimum lot size regulations. Typically if the home is much narrower than the minimum lot width, we will ask for a reduction in minimums that makes more sense, while maintaining the intent of the lot square footage in the zoning classification, by having the average lot size exceed the minimums.

Figure 8.26 Cantura Coves in Mesquite, Texas

Chapter Eight: Coving 101

Figure 8.27

Figure 8.28: The Villages of Creekside, Sauk Rapids, Minnesota.

San Cristobal Village

The original approved conceptual plan for San Cristobal Village in Santa Fe County, New Mexico, follows all the precepts of New Urbanism. Figure 8.29 shows the Northwest Section of the original master plan of San Cristobal Village. The brown area on top was an approved extra 500-foot setback to screen the neighboring large lot homes (comments on this type of absurd negotiation tactics with those opposing a development are in Chapter 9). The pink objects represent drainage washes. The gray areas represents high density mixed-use and commercial. The street pattern serve as entry and exit points to the 2,700 families who will pass through making approximately 15,000 trips a day!

We were approached by DR. Horton to do a coved layout for this site. The revised site plan (Figures 8.30 and 8.31) takes a totally different direction harnessing elements of Prefurbia. Instead of using the street pattern also as the walking pattern, the Prefurbia plan separates the two systems. The New Urban plan's parks were islands surrounded by streets, forcing those pedestrians to conflict with the 15,000 vehicles driving by each day, the Prefurbia plan reduces those conflicts.

The dark brown walks are the major interlinks with minor walk traffic shown in white. The different color homes represent a change in target home market. Some of the homes are served by

Figure 8.29: Original approved New Urban site plan.

wide rear drives instead of narrow alleys – adding to the variety. In Prefurbia, alleys can be accomodated, if need be. The constant change in form and shape allows each person traveling through to have a unique visual experience as they pass through the three square mile development!

The bulk of the traffic is handled on a collector system with decreased traffic along the internal residential hamlets. The coved neighborhood may appear less dense. It is not. In fact, the coved plan demonstrated an increase in density of 300 units from the original, along with an increase in average lot size of over 1,000 square feet!

All of this was accomplished while increasing local park areas by over 250 percent compared to the New Urban plan. On very large neighborhoods with thousands of homes such as this one, it is difficult to design patterns and maintain a sense of uniqueness throughout the site - but we tried to vary the streetscape and elements to create unique experiences and focal points.

The bottom line

Proper coved design has been demonstrated to reduce the length of street compared to conventional design between 15 and 45 percent. Assuming a reduction on average of 25 percent, this represents a corresponding reduction of street, walks, sewer pipes and services. But that cost varies widely throughout the country, constantly changing as economic conditions, materials, fuel, and regulations affect construction costs.

Some may argue that the increased driveway length negates the advantages of street reduction. That would not quite be correct, as driveways per square foot will always be a small fraction of the costs of public street construction, and the increased volume of driveway is very small, especially when care is given to taper to a narrow width on longer drives.

Taking the theory of Prefurbia to a higher level of detail is the Art Driveway shown (Fig. 8-30). Borrowing from turn of the century driveway design, we present the driveway as a sculpted artwork with a fraction of the environmental impact (and costs) of a standard driveway.

Chapter Eight: Coving 103

Figure 8.30 and 8.31: The revised coved plan separates the walking and street plans while increasing local park areas by over 250 percent and lot size by over 1,000 square feet.

Vitamin G - the human need for green

In an article published in Informed Design (informedesign.umn.edu) by Frances E. Kuo, Ph D., and Eric Miller, they relate the human needs for Vitamin G (green). Studies show that humans are 'phytotrophic', - people are attracted to environments that include trees, grasses, and other natual elements. Their report suggests that environments lacking Vitamin G lead to psychological breakdown in individuals. This can result in poor impulse control and increases in aggression and violence. Studies show that regardless of income, findings are similar - both rich and poor and everyone in-between need Vitamin G.

Coving reduces the amount of street required to attain a certain target density, increasing greenspace. The by-product of this design technique is that Vitamin G is not a 5 or 10 minute walk away, but adjacent to most (if not all) homes.

Prefurbia provides healthy doses of Vitamin G!

Other movements such as Landscape Urbanism taught in Harvards School of Design closely parrot the values and goals of Prefurbia. The methods introducd in this book are proven in the marketplace. With over 1,100 neighborhoods designed by our studio alone contracted by over 300 developers in 47 States and 18 Countries Prefurbia is indeed a validated solution.

Conclusion

Builders who have built in properly designed and executed coved neighborhoods tell us they enjoy faster absorption rates (sales of new homes) than in previous experience in conventional suburban subdivisions

Figure 8.32: Using more organic design we reduce costs and environmental impacts while increasing appeal.

CHAPTER NINE
Mixed-Use and Multi-Family Housing

"Any city however small, is in fact divided into two, one the city of the poor, the other of the rich. These are at war with one another".

— Plato circa 400 BC

A New Urban solution for suburban settings places residential units above commercial buildings. High density apartments and condominium units have the convenience of services on the lower level(s). These New Urbanists assume there is no need (or desire) for residents to have a car since public transportation hubs are a short walk away, and the same should apply in the suburbs. Another assumption is that residents actually desire a setting above active commercial uses that overlook vast areas of paving, traffic and parked cars. When these assumptions include families' as occupants, one has to question whether a consensus of opinions from occupants was gathered.

Figure 9.1 A town center east of the Addison Airport in the north Dallas, Texas, metro-plex.

Ideally, the shops and services on the street level serve customers outside the immediate area. The 'street-level' orientation of the architecture is also known as 'human scale' design.

These housing solutions typically found in urban environments cater primarily to two markets: Professionals without children and with busy lifestyles who find urban dwelling and its associated night life near the office exciting, and empty nesters who may have outgrown the suburban areas they migrated to raise their children. They may desire to be in an environment where restaurants and shopping are a few blocks away.

What works in dense urban settings may not work in suburbia. Having suburban residential on top of commercial is a risky venture unless there is abundant park space instead of paved spaces for quality views and a sense of serenity. Residential units above commercial property provide a poor quality of life for families.

In suburbia, residents may live in 'bedroom communities' many served by a regional mall. If there are localized commercial and office uses they are typically set in a 'strip-mall' design.

Strip Mall Advantages - ugly as they may be

From a retailer's standpoint, when shops are located along traffic routes, the exposure attracts customers, as shown in Figure 9.2 of a strip mall in St. Michael, Minnesota.

In a mature suburban setting, a coffee shop would best be located on the outgoing traffic side of an arterial street for morning commuters, inviting a quick stop on the way to work.

The strip mall is designed for convenience. It allows the driver to recognize there is a dry cleaner, then easily turn into the center and park in front of the store, drop off their items, and be on their way.

The exposure and easy access is what makes strip malls so desirable. The customer gets their product and retailers make their profit!

Figure 9.2 A strip mall in St.Michael, Minnesota with adjacent senior housing.

Exposed Rear

The back side of the typical strip mall lacks architectural detail and is used for waste disposal, employee parking, and loading. If not screened properly, it is an eyesore to surrounding residents.

Proper screening is very expensive, as can be seen in Figure 9.3 with the tall brick wall in the rear of this California strip mall set against single family homes. That is – 'if' there is any screening.

Figure 9.3 The loading docks with a wall to screen the adjacent single family homes

Let's revisit the strip mall in the previous aerial photograph from the view point of the senior housing units behind the strip center:

This (Figure 9.4) is not a great view to come home to. This wide public street separates residential and commercial property.

Figure 9.4 The strip mall in St. Michael, Minnesota with no visual screening.

Plain featureless store rears, loading docks, garbage bins and associated aromas are not quite the experience the senior residents across the street were hoping for at their retirement age.

Why is there a public street along the rear of this center in the first place? Later in this chapter we will present more efficient alternatives, the Neighborhood Marketplace.

Making matters worse, residents on the other side of the street often have this view (Figure 9.5)!

Figure 9.5 The other side of the above truck.

Figure 9.6 This is the welcoming view into a suburban city.

Commercial Architectural Control?

Strip malls are often at the front door of the community in which they serve. Figure 9.6 shows the view upon entering the city of East Bethel, Minnesota (circa 2006). What is your impression?

Nationally, the façade of the typical suburban strip mall is rarely up to the architectural standards that present the town as a desirable place to live. Often the first impression one gets is a mish-mash of cheap looking structures, seas of asphalt, parked cars and trucks, a variety of signage competing for attention - and a lack of landscaping.

This low standard is not good for the city or its values. This is nothing new. Virtually all books on suburban retail explain how bad this situation is. However, few cover the next problem.

Where is Commercial exposure?

Many strip malls are designed in a layered 'outlot' fashion. The developer initially builds the strip mall deeply set back from the arterial street, to leave room for the future retail (fast food) outlots.

After the strip mall is leased and established, the outlots are sold to fast food chains, whose presence when built shields the view of the original strip business' from customers, eventually choking them out. Designing strip malls for failure is *unsustainable*!

The developer walks away with huge profits, leaving those that leased in the original center struggling to survive. The visual clutter of the outlot commercial also tarnishes the image of the city. Restaurant rears of the outlot are typically exposed, further destroying the character of the shopping center and town.

What gets placed behind commercial loading docks? Multi-family housing! Yes, it seems rational thinking is to place the maximum number of people along the rear of these strip centers in higher density multi-story buildings to overlook the screening walls onto the ugly side of strip malls. How much sense does that make? How 'balanced' is that?

Where are Commercial services located?

The key to success for commercial property is location, location, location. This means two things: exposure and customers. If the location has plenty of customers (enough population density to support the business) but is located where no one can see it – the business will likely fail.

That is why it is absurd to place retail within a low density suburban neighborhood.

Another reason commercial services may be poorly located, is that most residential developers in suburbia are just that – residential. A residential developer is not likely to have commercial expertise. They understand a two story home, and not a Dairy Queen.

A home builder that purchases 300 acres usually wants only houses that he can sell on the entire 300 acres. The residential builder often thinks that commercial development is just a blemish that can diminish the homes' values!

The Neighborhood Marketplace

What if you can have all the advantages of the strip mall – without disadvantages and actually increase the value of both retail and residential property?

Many suburban commercial strip malls have businesses that can survive without loading docks. The area of paving and screening walls, along with associated extra residential setbacks, can be transferred to better architectural detailing that enhances the community and promote business.

Figure 9.7 This is the "rear" of a strip mall in Charlotte, North Carolina.

The logic for a change

Assuming in suburbia that the strip mall is adjacent to residential areas, that strip mall can be transformed into: The Neighborhood Marketplace.

Figure 9.7 shows the rear of a strip mall in Charlotte, North Carolina. Yes, that is the rear! Notice nobody is parked back there. Everyone has parked in the larger parking areas in the front of the stores. I am taking the photo while standing next to a 10-foot high stone 'screening wall' about an eighth of a mile long. This wall screens the view from the adjacent townhomes.

If someone living in one of the townhomes wants to get a pizza at this shopping center, they must get into their car and drive about a half mile to go to the restaurant that is just a few hundred feet away!

The view of the rear of this center is much nicer than looking across the townhome private drive to the row of garage doors across the way.

If the wall was removed, as well as the parking that no one was using, that land could have been used for detention ponds, fountains, board walks, patios...
...and quite a few more townhome units! The internal walk system could lead all residents to the shops giving them an excuse to walk instead of drive!

Figure 9.8 A plan that includes The Neighborhood Marketplace.

The concept of The Neighborhood Marketplace relies on merging planning and architecture. From an architectural perspective, Tony Dellicolli, AIA, at Cityscape Architects Inc., in Novi, Michigan, provides the following insights:

"The key to marrying architecture and planning together for The Neighborhood Marketplace is the scale and the level of execution of architectural detailing.

It is important for developers and owners to recognize early in the process how important it is to develop an architectural style that can be executed throughout the development.

The style or theme of the residences needs to be carried out in the architecture of the Neighborhood Marketplace. The architecture must have a residential flavor reflecting the same level of detail as the residential homes".

This level of detail can be seen by the aerial photograph, Figure 9.9 of the commercial center at Liberty on the Lake in Stillwater, Minnesota. Liberty on the Lake gets the architecture correct - stores with no loading docks, stores that have exposure to arterial roads. Unlike the concept of The Neighborhood Marketplace, streets and parking disconnect the retail from the residential - so it does not fall into The Neighborhood Marketplace category of design.

More often than not, the typical architecturally-detailed retail strip center is designed to look like a false façade or Hollywood set, where the customer is looking at a structure built only for the purpose of reducing commercial impact – note the rear of the roofs. Often, these obvious shortcuts cheapen the neighborhood and/or limit functionality of the intended uses.

In Prefurbia, retail stores must be designed for good exposure to arterial thoroughfares.

For example, the store front needs to provide easy access for the vehicular customer, and at the same time provide for the patrons visiting the store on bike and foot. The space between residential and retail is part of an overall connective open space system within a larger neighborhood. It is the control of flow between residential and commercial uses with both pedestrian and vehicular systems that make the Neighborhood Marketplace work well.

The open space between residential and commercial becomes a great place to locate decorative detention-retention ponds, boardwalks, patios, etc.

In some situations there may be no way around having all stores serviced from the rear. This is where inventive architectural solutions are necessary.

The rear of these centers can be cleaned up visually by simply placing the loading zones into hidden service courts carefully situated between the retail shops. By the use of masonry screen walls and landscaping, access to the stores and service delivery areas can be totally hidden, as shown in Tamarack Village, a Woodbury, Minnesota, retail center (Figure 9.10).

Figure 9.9 The commercial buildings shown here are surrounded by parking.

Chapter Nine: Mixed-Use and Multi-Family Housing 113

Figure 9.10 Tamarack Village in Woodbury, Minnesota, has some wonderful features, however, walkability is not one of them.

Not many people are aware of where the delivery services are placed for all of the stores in a regional mall.

This concept of The Neighborhood Marketplace will allow the developers and municipalities to eliminate the need for any type of tall, opaque screen walls to separate the residential from the retail strip center.

Traditional Land Use Transitions

Figure 9.11 is a typical example of development that intermixes different types of housing, a common suburban pattern.

Most people would find nothing wrong with this plan. The project is better than the typical cookie-cutter subdivision. Let's take a closer look at this common scenario.

With transitional zoning, land uses, (typically equating to changes in density and price points), transition from lower to higher homes values.

Let's look at the entrance to this development:

The highest density – lowest priced homes are

Figure 9.11 The most common suburban transitional land use - low priced housing progressing to high priced.

Figure 9.12 The lowest priced housing is at the entrance.

Figure 9.13 Price points get higher as one drives through the development, and residents that can afford it met more space.

Figure 9.14 Placing the most families with the worst views is not a sustainable solution.

set along major traffic corridors of this suburban town.

For those new to the area or unfamiliar with the city, what impression will they have of the area? Development (residential and commercial) has a tremendous influence on the long term, economic stability of a city. Cheap and dense housing along corridors does not promote a sense of arrival.

With the exception of five units in the shaded red zone (Figure 9.12), no other units have open space adjacent to them. It's a sea of pavement and rooftops.

Outside the red zone are more townhomes, but they are in a row pattern along open space and in the next higher price point.

The next form of transition is single-family detatched townhomes. Most of these homes have some view of open space. They represent the jump in price point and transition from attached housing to detached single-family lots. Figure 9.13 clearly shows the transition lines.

If you think in terms of housing as being a product, how does driving through lower-priced, high-density homes make the higher priced, lower-density homes down the road more valuable?

Over the years as homes sell and re-sell, how will this transition help preserve home values? It doesn't, because now that the city has achieved density, the property adjacent to the lowest priced homes will likely be commercial and loading docks. (Figure 9.14) Highly detailed-planning a region can help solve zoning transition problems. Instead of a broad-brush look at a region with 10 acres of townhomes here and 20 acres of single-family homes there, a more in-depth look at what makes *design-sense* is needed.

Transitions that Preserve Neighborhood Quality

Some designers think intermixing price points and throwing away transitions altogether should be the solution.

In that scenario, the poor live next door to the rich, all living in harmony. There is no crime and people of different races and religions embrace each other.

This dream scenario hasn't worked yet. Juxtaposing low cost homes and expensive homes will have a positive effect on the values of the lower priced homes, but a disastrous effect on the high priced home values. Even the Congress of New Urbanism which promotes a utopia of mixed income make note of their 'gentrified' developments success rate. Gentrified is not mixed income.

Transitions, as evil as they seem, somewhat preserve property values. That is the main reason they exist. The problem is that we have been transitioning the wrong way!

Why not have the quality housing along the arterial streets that create a sense of arrival, and then transition to affordable housing? This can be accomplished in a variety of ways.

Connective Neighborhood Design (CND)

An organized approach to reverse the commonplace transitional zoning in Prefurbia is called Connective Neighborhood Design, or CND.

This is best explained using an actual before and after case study of Westridge Village in Eau Claire, Wisconsin. Westridge Village was designed and approved by the city as a Planned Unit Development (PUD) to intermix different housing types and add commercial property along a major road into the town. The site was then picked up by RHS Development, who researched options to create a better neighborhood model. RHS chose Connective Neighborhood Design as their solution.

Westridge Village — A case study

Figure 9.16 represents what could be considered a very typical suburban planned unit development

To help you understand what is going on around the site (north is up):

Figure 9.15 A scenic view in Westridge Village

Along the north border is Highway 12 – a major road that leads to the city of Eau Claire. The north side of the street opposite of the neighborhood is industrial.
To the west are older homes that can best be described as average.
To the east is an arterial street, with little high traffic and a view of rears of townhomes across the street.
To the south is Interstate 94 connecting Chicago with Minneapolis-St. Paul.
Now, let's examine the plan design flaws.

1) Streets to that lack   destination

There are seven major streets (shown in different colors in Figure 9.17) all ending in a cul-de-sac or T- intersection. The straight streets encourage speeding, make more dangerous because of the abundant 18 total intersections. In all, 32 acres of land are consumed by street right of way.

2) The main park is far away

Figure 9.16

The parks for the original approved Westridge Village are indicated by the large green areas in Figure 9.18. You may wonder why the park is along the west edge of the development, and not central for all residents to enjoy? This location was chosen to appease the existing neighbors and gain approvals with less opposition.

This means that the 500 families, will forever suffer by being forced to take a long hike to the park (it will then be underutilized) so that the 20 existing residents adjacent to the west property line will support the project! What is even more absurd is that every week across the nation, subdivisions are designed and approved under this very scenario! Keep in mind that, statistically, the typical home sells once every six years and those neighbors that initially fought the development are likely to move before the project is completed!

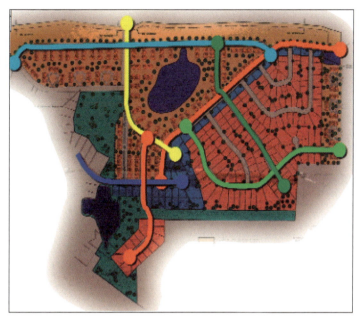

Figure 9.17 There is no main street - all streets designed with no apparent system.

3) Walks with no particular destination

There is no provision for walks other than the very-difficult-to-traverse street system, encouraging residents to use their cars resources, increasing pollution as well as the waistline.

4) Prominently displays multi-family housing

The only single-family detached housing (shown in red on Figure 9.19) is well hidden behind the blue duplex lots. Consider this; if you are not visiting someone in the single family homes, the entire development would appear as a cookie-cutter multi-family project. Making matters worse, only one type of multi-family unit (a ROW-type 4 plex) is indicated, eliminating the ability to have a wide variety of affordable housing types.

Figure 9.18

Worse yet, this scenario (as indicated by the dollar signs) showcases only the lowest cost housing and hides the higher priced homes, typical of the transitional zoning that suburban towns across the USA widely embrace.

5) Highlights unsightly rear-yard clutter and plain rear architecture

The view along the arterial and internal streets, except the single family areas, is of side and rear yards. In the Midwest, fencing is not the norm, so yard clutter would reduce the quality of this neighborhood as it ages. Also typical is one-sided architecture, which is reserved for the front of the homes. This plan would enforce the mundane vinyl-town look common with today's suburbs.

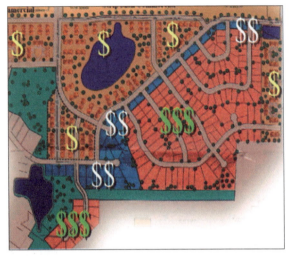

Figure 9.19

Westridge Village as a CND:

By considering at the onset, the details and elements of designing a neighborhood, from walking connectivity, views from within the homes and along the streets, from grading, utilities, etc, it is possible to create a better neighborhood (Figure 9.21).

Let's look at Westridge Village from the perspective of the points mentioned earlier.

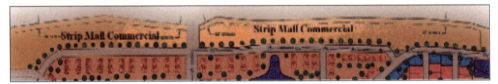

Figure 9.20

Safety layer

Areas where pedestrians and bicyclists intermingle with vehicles, accidents are common. These systems have been separated as much as possible throughout the entire neighborhood.

There are no speed-inducing streets and the number of street intersections have been reduced by half. There are walks going through blocks and at ends of cul-de-sacs. These can be utilized by emergency vehicles in the unlikely event that a street is blocked.

Environmental layer

Looking at Figure 9.22, the landscaped islands are used for aesthetic appeal and detention of stormwater, creating a dual use of the space. Surface flow through the neighborhood to the man-

Figure 9.21

made lake reduces the need for traditional sewer pipe, thus there is some absorption of run-off that would not otherwise be possible with a strict piped design stormwater system.

The land is also used more efficiently, as this revised plan increased density by 126 units!

Economic layer

With 126 more homes on the same site, affordabiliy is improved

Just for assumptions, let's apply some numbers:

The reduction of infrastructure costs (due to the more efficient design) allowed the addition of the extensive walk system while staying within budget. The Neighborhood Marketplace proposed at the entrance will attract customers entering off of Highway 12, as well as the pedestrian traffic within the neighborhood.

Figure 9.22

Aesthetic layer

The freeform streets and walks, meandering setbacks, and the increased openness will make this a special place for all residents, regardless of income level or social status.

Loft Units:	$120,000	Twinhomes:	$180,000
Townhomes:	$170,000	Single Family:	$250,000

The number of units would be:

	Conventional	CND
Townhomes	332	270
Twinhomes (Duplex)	70	32
Single Family	172	344
Loft Units	0	54
Total	574	700

The gross dollars generated would be:

	Conventional	CND
Townhomes	56,440,000	45,900,000
Twinhomes (Duplex)	12,600,000	5,760,000
Single Family	43,000,000	86,000,000
Loft Units	0	6,480,000
Total	112,040,000	144,140,000

No home rears are visible as one drives or walks through the neighborhood. The beautiful lake is ringed by a walkway for all to enjoy, not just an elite few.

There are no visible lines of transition. Low, medium, and high-end housing is not defined by left or right of a noticeable transition line, which aids in creating harmony when this neighborhood is complete.

Architectural and landscape layer

The larger number of units allow architectural and landscaping features to be standard and still be within economic reach of the typical Eau Claire family.

The continuity of architecture throughout the neighborhood will blend economic classes.

The planning of this site would be next to impossible with a broad brush approach. Without looking at each and every detail, including the placement of the homes, walks, the drainage, and the traffic patterns, etc., this sustainable development would be hard to replicate.

Prefurbia Multi-family housing

Discussion of multi-family housing can easily consume a separate book. We will instead introduce better concepts to create functional attached housing development. There are two distinct markets utilizing multi-family housing, renters and owner-occupied.

In the rental market, which many consider to be serving a transient population, supplying affordable housing is a primary goal. Too often, affordable units are positioned as noise and site buffers along highways, in back of factories, etc. It is as if 'those people' don't need quiet nights or quality views. The recession taught us we are all just one bad investment away from living in a low income rental unit.

Owner-occupied housing is typically comprised of townhomes and condominiums. As with rentals, too often these units are placed as noise and site buffers along highways, in back of factories, etc., as if 'those people' don't require quiet nights or great views.

Multi-family housing is often treated (in design) as if the residents are second-class citizens. Earlier we covered better placement alternatives for multi-family housing in a mixed-use development.

Figure 9.23 illustrates a townhome neighborhood that uses the same attention in its design as airplane hangars. Aircraft don't care about the views out of hangar windows, but people do. Of the hundreds of people who will live in the townhome development, only a very small percentage will ever be aware of the large open space along the adjacent wetlands during their daily lives. The next chapter introduces Architectual Blending to enhance the quality and living standards of multi-family development.

Figure 9.23: Hangar row design that is all too common.

Minimize development impact

Run the numbers. Suppose raw land costs are $80,000 an acre and the density achieved is 14 homes per acre. This results in a 'land cost per unit' of $5,714. If density is reduced to eight homes per acre to create a less dense neighborhood, the cost increases to $10,000 per unit. However, look at the wide public street in Figure 9.23 that provides little frontage – this is a wasteful design. If the streets were more efficiently designed, then perhaps the extra $4,300 per unit would have been saved in paving and utility reduction. In other words, better design can leave the developer building a beautiful neighborhood instead of a barracks-like project, while preserving profits.

To see how this works, look at this study using a 'before' plan in Ypsilanti, Michigan, Figure 9.24, that was eventually redesigned. The typical developer would praise the designer for jamming as many homes on this site as possible. No doubt the design fills all of the available area. Unfortunately for everyone, the 'fill' is mostly paved surfaces, not housing units.

It is very easy to lose track of design efficiency, something we are about to teach you.

From the plan shown on Figure 9.25 you will see a different picture.

The grey area represents the area of public street where driveways could be allowed to be placed.

The red areas represent paving which serves no home frontage – no real reason it should be there. The yellow area is half-utilized paving (fronts of homes on only one side of the street).

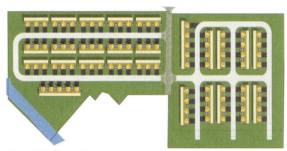

Figure 9.24 The original design

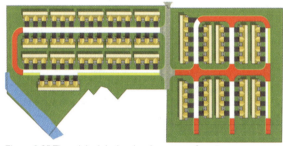

Figure 9.25 The original designshowing areas of waste.

Of the original paved area, this represents a 72 percent loss of efficiency. Bet you did not see that when you first looked at the design. That excess of *over an acre of waste* can be utilized for open space and density gains.

Is the above plan a project that will provide a high standard of living for residents to influence them to make this their long term home? Is this a design that becomes a welcome asset to the surrounding community? Most importantly, is this a design that elevates the living standards of its residents?

Again, sustainable development needs to work on achieving balance. This includes a desirable quality of living for multi-family residents. It is even more important to create a neighborhood that residents will continue to want to live in, one that fosters a sense of pride. Now let's look at the revised plan (Figure 9.26).

This plan uses the same unit foot print, only it has been modified to have fewer units in a straight row, and has a small offset between pairs of homes, to create variety. The resulting neighborhood would be more inviting to residents and be a place where everyday living would preserve resident's dignity. Only a few homes are placed close rear-to-rear.

As far as density, the original plan had 179 units, the re-planned one has 186, a gain of 7 homes. How is this possible? First, seek out the waste in the design and then eliminate it while using the extra area to create something special. Eliminating waste is not taught in urban planning courses. However, we cover it in this book and at length in the LandMentor system.

This multi-family neighborhood (Figure 9.27) in Albuquerque, New Mexico, has variation within the placement of each unit which achieves a panoramic view from within the home of most of the units. Obviously this is an upscale development, but it would not take excessive costs to emulate this design at any home

Figure 9.26 The more livable design after reducing waste

Figure 9.27 Staggering townhomes allows increased views.　　Figure 9.28 These units are staggered but no side views.

price point. This creates interior units to have the same advantages of the premium 'end' units, which allows for more window area.

Boulder Ridge, in River Falls, Wisconsin, Figure 9.28, clearly shows the extreme staggering of the units to allow for broad panoramic views. It would have been even nicer had the builder actually placed windows on the units 'sides'! The developer never told the builder that the site plan staggered the units to enhance views from living areas, and the city did not seem to catch this either! In the next chapter we demonstrate how to use architectural blending to enhance value and quality of living at all price points.

Doing the right thing

Leo Sands, a hotel builder from St. Cloud, Minnesota, commissioned us to design a low income subsidized multi-family development in Mahtomedi, Minnesota (Figure 9.29).

Why is there such nice architecture with every home having a huge porch overlooking a wooded wetland? Perhaps Leo did not follow conventional thinking, that 'those people' should be located along loading docks and stoops.

This is a multi-family neighborhood that achieves a balance of the many layers referred to earlier, a sustainable neighborhood. This housing is at the entrance of a larger developed area which includes exclusive townhomes, few of which look this nice!

But not all multi-family housing is built because it's cheaper than single family. In fact, many townhome developments are built because there are a growing number of people with means that simply do not want to maintain a yard. Many 'townhomes' cost more than a single family home of similar size, in some cases much more.

Figure 9.29 Low income housing done right

Staggering housing units reduces monotony. Other methods include twisting units so they are not parallel, so side views can be expanded reducing residents viewing directly into each other's units.

Enough of a twist may allow for a side entry garage or even some three-car townhomes.

In Villages on the Rum (located on the Rum River) in Isanti, Minnesota, the preliminary plat included elevations of the buildings along the entrance. These were to be the 'affordable' units at low price points. We insisted that the developer have architecture set in stone at the preliminary plat stages, eliminating bait and switch tactics.

Note the huge setbacks which signal that the neighborhood one is about to enter, offers an embracing sense of tranquility and space (compared to other developments in Isanti).

Figures 9.30 and 9.31: Villages on the Rum

SECTION FOUR

Advancing Architecture
Prefurbia for Redevelopment
Model Regulations
Technology & Education

- BayHomes & Architectural Blending
- Prefurbia for Redevelopment
- A Model 'Rewards Based' Cove Ordinance
- Technology & Education

CHAPTER TEN

BayHomes

"New Urbanism with a view."

— Patrick L. O'Toole

Senior Editor, *Professional Builder* magazine explaining BayHomes

(January 2001)

    The 'BayHome' (above, left) represents a model for single family housing which utilizes townhome zoning. The major differences between a 'detached townhome' and a BayHome is the placement of home fronts along common open space (instead of fronting a street) and the BayHome interior architecture that *blends* its 'living space' with adjacent exterior 'open space'. They are typically a more affordable option for single family living. BayHome design requires specific home interior layout, defined room functions, wall and window locations - all of which become elements of the overall neighborhood design. A full 'useable' front porch and connective walk system encourage residents to enjoy the outdoors and foster neighborly interaction. The BayHome led to many design innovations introduced in this chapter.

    By individually orienting homes and specifying floor plans, it guarantees panoramic views from most, if not all, units.

Similar to TND or New Urban architecture, a BayHome lot is narrow and deep - with a mandatory front porch. Garages are situated in the rear, but unlike New Urbanism, they are attached (Figure 10.1).

The 'Bay' in the name BayHome has little to do with the home unit. It refers to the shape of common open space. 'BayHome' was chosen because the open spaces often resemble shapes of 'bays of a lake' as shown in Figure 10.2.

Figure 10.3 of the same image removes exterior walls exposing the first floor interior walls. Functions of individual rooms become a main 'site plan' component. The interior gathering place for the American family is the kitchen. This 'focal point' opens up through living areas that also leads to the front porch - which connects to neighbors along walks (pink) through commons.

BayHomes were the first land planning method where interior and exterior spaces are 'blended' in *production* housing.

All BayHomes in this example offer panoramic views from the kitchen onto the commons (Figure 10.4). Homes have windows that wrap along front and sides of the homes.

To achieve panoramic views and blending of spaces, land planners and architects must collaborate at the initial stages of site planning.

The 'staggering' of the home avoids views into the neighbor's house.

Figure 10.1 BayHomes have attached garages

Figure 10.2 The shape of open spaces often create a 'bay-like' shape.

Figure 10.3 It is the management of the interior functions, wall locations and window placement in concert with the exterior spaces that made BayHomes unique. All of this requires an attention to detail that consumes just a few minutes extra effort for each unit to create value that can last for centuries.

Figure 10.5 is a view from the kitchen counter into a portion of the living space in a million dollar, 28' wide BayHome in Grand Haven, Michigan. The open view and light that enters the home is unequalled by any of the association-maintained homes in the region - at any price point.

The majority of BayHomes have a first floor master bedroom to serve a broader housing market. A first floor master bedroom could accomodate adult, senior, or semi-assisted-living residents.

Figure 10.4 Panoramic views of vast commons from the kitchen are provided in these BayHomes of Remington Coves. Note the corner windows of the right side of the home, except the red unit where the land surveyor forgot to 'flip' the floor plan! An addition to attention of detail required on plan must also translate to the field during construction! Thus, all methods in Prefurbia must also be communicated to everyone after plan approval from the construction crews to the sales people - all participate in the overall success or failure for developers and builders and ultimately city.

Solutions for planning and land use-related issues

Homes along arterial roads

As stated earlier, a major problem with residential planning is how to design homes at the edge of a property, especially along the arterial streets, where rear yards are often exposed and tend to clutter, or contain fences or screening walls (which adds several thousands of dollars to the price of every home).

Positioning BayHomes along arterial streets with deep setbacks (figure 10.4), the window wrap, staggering, front porches, meandering walks, and landscaping creates a 'village-like' welcoming for a neighborhood. This would encourage potential home buyers to 'check out' the neighborhood instead of just passing it by. It is a better alternative than attached housing, screening walls, or wasteful earth-berms.

Figure 10.5: A million dollar BayHome designed by Doug DeHaan

Placement of BayHomes along higher volume trafficways, or undesirable views such as industrial or treatment ponds has additional advantages. Since the rear of the BayHome is garage, it provides a good noise barrier. The rear of the architectural design places services along less desirable areas, leaving the homes to face quality views to premium locations. If more screening or noise buffering is needed, the rear drives can also service mini-storage units for the neighborhood. Thus, screening undesirable property while serving an important need for the overall development.

BayHomes are also ideal along high visibility and high value edges of the site, such as golf courses, wetlands and water edges (Figure 10.6). The walks along the BayHome fronts invite the entire neighborhood to take a stroll and enjoy the views.

Figure 10.6 BayHomes at Greenbriar Hills, Buffalo, MN along an arterial street shown in a 3D created by LandMentor.

Certainly when golfing or boating, the view of porches and landscaping is nicer than looking at rears of residential units (Figure 10.7).

Affordable housing

There are no particular 'price point' targets for BayHomes. They can be a solution for lower, middle, as well as the upper income home buyers. There are no maximum unit widths. As a guideline for middle to upper-middle-income housing the majority have been in the range of 22 to 32 feet wide. A two story 24-foot wide BayHome can easily provide over 2,000 square feet of living space. If there is a finished lower level (basement) and living area over the garage, the same 24-foot width can exceed 3,500 square feet of living space.

A common element of *all* BayHomes is that the private drives replace expensive (and excessive) streets, typical of single family 'lot' development, lowering construction costs. When affordability is an issue, the home can be narrow, however, for two-car garages the practical width limit is 20 feet.

Another BayHome advantage is the yards are association controlled and maintained. This also prevents yard clutter. Since some lower-income families struggle to maintain their vehicles in optimum condition, their cars are placed behind the homes,

Figure 10.7 BayHomes proposed along a golf course, Ramsey, Minnesota. Architecture shown by DeHaan Architects

generally out of sight, thus eliminating vehicular clutter that would otherwise devalue a more traditional gridded neighborhood.

As far as the monthly maintenance fee associated with common areas such as the gardens and the care of the private drives, these are factors in any form of home ownership. If the owner of a BayHome cannot afford the maintenance fee, they are not likely able to maintain a single-family or townhome either. Home owners, whether rich or poor, are the stewards of the community and must maintain their property at some basic level for stability of the region. BayHomes provide affordable housing that can be a positive showcase for the community (Figure 10.8).

BayHome density advantage

BayHomes are placed 10-feet apart to achieve density that renders twinhome (duplexes) and much of the townhome market obsolete (because density is similar). While BayHome density has been demonstrated up to seven units per acre (with a 22-foot wide product), they more commonly achieve four to six homes per acre while delivering a sense of space one would expect in a more typical large lot suburban single family neighborhood.

At greater than six homes per acre, attached housing may be a better option. Is it the best financial decision? From a developer's standpoint, attached multi-family 'lots' sell for much less than a single family home lot. Using a hypothetical scenario, let's suppose a developer purchases 40 acres at $50,000 an acre. The same developer is proposing to rezone the site to 20 acres of single family 10,000 square foot lots and the remaining for multi-family at 10 units per acre. The resulting 60 single-family lots may sell for $70,000 each and the 200 townhome lots sell at $25,000 each.

The developer's gross revenue is $9.2 million [(70,000 X 60) + (25,000 X 200)].

Figure 10.8 BayHomes in The Villages at Creekside, Sauk Rapids, Minnesota

Potential buyers of the 60 single-family homes will look at the adjacent 'hangar row' layout of the townhomes, as having a negative influence on their buying decision. The builder must also move (sell) 200 townhome units, which could take a long time.

From our tracking of existing coved neighborhoods which included BayHome lots, we have found that BayHome lots sell for close to single-family lots, about 80 percent of the single-family home lot price. This means that if the townhomes were instead BayHomes, those lots would sell for approximately $56,000 (80 percent of $70,000). A reasonable density would be 5.5 homes per acre, or 110 BayHomes on the 20 acres.

Re-running the financials, if the 20 acres were BayHomes, the developer's gross revenue is $10.36 million [(70,000 X 60) + (56,000 X 110)]. There is over a million dollar advantage to the developer selling these individual lots to builders, not to mention the preferred quality of life.

With BayHomes, the builder has a market advantage over attached housing. Even though BayHomes will need to sell for a higher price in this scenario, they can still undercut single family homes of similar square footage on an owner-maintained lot. Another advantage is that there are fewer homes to sell, reducing the time a builder is exposed.

BayHome – chicken or egg scenario

A big hurdle in BayHome planning is obtaining the final architecture and floor plan. Ideally, the architecture should be set before the initial planning of a BayHome site. To design initial concept plans may not be practical without knowing the direction of the architecture.

In cases where the builder is also the developer, final design decisions can be made fast. If it's the developer selling to a builder or a few builders, that final design decisions may take longer. However this situation is not unusual when a developer desires to create townhome or multifamily projects, those also require a good idea as to the final architecture at initial site planning stages.

The BayHome land planner could make up 'mules' or test units that have a generic layout of the rooms and overall dimensions to get the feel and density numbers, and later fine tune the design before final plat approval. In this scenario, both planner and architect must be involved from preliminary platting, all the way to the final stakeout!

Another solution is purchase existing BayHome designs before any planning is done. This way the preliminary plat can set both lot areas and interior floor plans simultaneously – thus guaranteeing the maximum advantage of the design. The architectural fees are typically passed onto the builder. Another advantage to this scenario is the developer could sell to multiple builders, yet maintain consistency.

Lot configuration

BayHome 'lots' can be configured in a variety of ways. Typically, lot lines extend five feet beyond the perimeter of a BayHome. Since the entire site is association maintained the lot is essentially has no real ownership advantage, unless it is a zero lot line with private side yard space. Illustrated in Figure 10.10 and Figure 10.1, are zero lot line BayHomes with private side yard spaces.

Private drive

BayHomes do not have narrow alleys, instead they utilize wider more usable private drives using dimensional standards under the cities existing townhome standards.

In no case should BayHome lane be less than 16 feet or they would become 'alley like'.

Cul-de-sacs (Figure10.9) or turn-arounds should be one way pavement from 12 to 16 feet wide. These are maintained by the home owners association. The private drives are not public streets nor are they typically constructed to the more expensive public street standards.

Garages

Another advantage of BayHomes is the inclusion of a three-car garage as an option. The staggering of the fronts needed for panormaic views mean the rears also stagger, allowing for many of the garages to be accessed from the side, thus significantly reducing the visual impact of a two or three stall garage(Figure 10.11).

Many BayHomes offer tandem garage depth, meaning that one side of the garage is two cars deep. The BayHomes in Remington Coves (Otsego, Minnesota) are designed this way. All appear to have two-car garages, but are in fact 3 car garages.

Why is a 3rd stall important? Many empty-nesters may want to have a house with association-maintained living, but refuse to sell their high maintenance single-family home because the alternate townhome rarely provides enough storage for their cars along with their boat, motorcycle, snowmobile, etc.

A full Front Porch

BayHomes require a large usable front porch (not a cosmetic 'stoop'). The BayHomes seen in the architecture from DeHaan Architects were used in many of the neighborhoods we planned (Figure 10-12). All have a large porch, with enough space to sit and socialize.

Without a full defined front porch, it's not a BayHome. Without walks connecting the homes and surrounding neighborhoods (if the walks continue) – they are not BayHomes.

A solution for families

At first it may appear that BayHomes are for the empty nesters because of the lack of a large private yard. Yet, there are plenty of *families*

Figure 10.9 The BayHomes 'plat' shown of Hillcrest in Altoona, Wisconsin. Approved in 2013, the BayHomes sold so well, the developer is expanding their reach through the development, previously a golf course.

Figure 10.10 BayHomes with private side yard space is make possible by using zero-lot line alternatives if allowed by local regulations. While adding architectural complexity to avoid neighbors looking into private space, it adds extra value without extra costs.

Figure 10.11 A BayHome with a three car garage

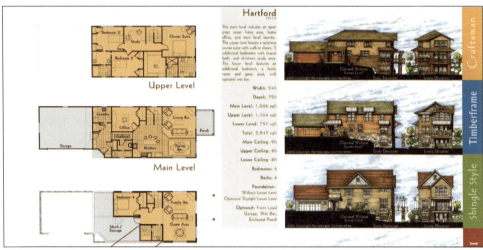

Figure 10.12 BayHome plans from DeHaan Architects

living in townhome developments that have no or little private yard area, and the BayHome is no different.

Private side yard space

Again, BayHomes can have private yards along the sides that overlap - not possible with attached housing. The plan in Figure 10.13 shows a BayHome neighborhood being built at Retama in Mission, Texas with private side yards, screened by fences.

This screened area would have no negative impact on the feel of the neighborhood open space. The extra design related effort is well worth it for increased value and livability.

Rear entrances

BayHomes have vehicular access at the rear of the home. While there are, in some cases, public streets in the front (porch side) where guests can park, in most cases guests will enter the rear of the home through the private drives.

Is this acceptable? BayHomes require four-sided architectural detailing to make the back of the home appear similar to a 'second front'.

If you think that is not acceptable, compare the typical *luxury* townhome fronts in Figure 10.14 and 10.15 with the single family detached townhomes.

Figure 10.16 show the fronts of upscale empty-nester detached, association-maintained homes in Maple Grove, Minnesota. There is no front door. Guests must enter living areas through a side entrance.

Figure 10.13 BayHomes with private side yards proposed in Mission, Texas at Retama Village by Rhodes Development

In the worst case scenario, the rear of a BayHome is no worse than any of these 'upscale' home fronts!

In the late 1990's the BayHomes represented a fresh planning approach. The ability to take production housing and design a site where interior and exterior spaces are planned at the same time, was initially unique to the BayHome.

BayHomes provided a higher quality of life for those seeking low maintenance living and where higher density is desired.

Before the recession we had thousands of BayHomes approved and many more on the drawing boards. The recession created a temporary halt to the demand which picked back up as markets recovered. Developers today seem to embrace BayHomes as much as they did before the recession.

During the downtime in the recession we began pioneering more housing options with advantages of BayHomes that would apply towards all forms of housing...

Figure 10.14 This 'luxury' townhome front is worse than any BayHome 'rear' entry area, yet acceptable to the home buyer!

Figure 10.15 This upscale townhome development in Woodbury, Minnesota - this is their front door!

Figure 10.16 Another upscale townhome development - side entry!

Beyond the BayHome

BayHomes gained momentum from 1999. Developers and builders slowly embraced the concept until much of our land planning included BayHomes. This experience provided a basis to add value to all forms of housing.

Evolution of Architectural Blending

As we developed LandMentor as a primary tool for the design of complex planning situations such as BayHomes, it became obvious the advantages that made BayHomes desirable could be applied to all housing types. *Architectural Blending*, the criteria that ties the interior of a residential unit into adjacent open spaces, influenced a revolution in design methods and the development of more advanced software technology. The existing simplistic design strategy, to concentrate on geometry to layout streets and then lots (ignoring the floor plan) becomes old school. The easy 'outside-in' method used in conventional design evolved into techniques to plan neighborhoods from the 'inside-out'. To design as if each home deserved the best possible views, as if the designer themselves lived in each unit. To accomplish this, we developed software technology that was accurate enough for engineering, yet easy enough for planners, landscape architects, and architects to utilize. At the beginning of this book we indicated how critical it was for planners to understand the needs of engineers and surveyors and conversely for engineers and surveyors to understand the importance of good planning. Architectural Blending requires that *architecture* be one of the main ingredients of the sustainable design recipe.

Architectural Blending creates a new standard

Many neighborhoods we design today integrate the home floor plan as a function of the neighborhood. Architecture becomes more important than adding a front porch, or 'craftsman' trim. The timing for this level of integration could not be better. Faced with an economy where new home buyers are facing higher prices and interest, builders and developers are finally realizing the old Unit 'A' with facade option 'D' is not going to be enough to leverage a sale. Competition demands that every market advantage be utilized.

BayHomes had already set the stage for the evolution of surrounding homes with space, providing viewsheds, and coving also delivers the surrounding space.

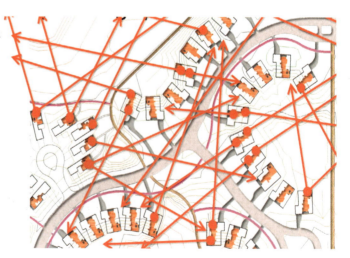

Figure 10.17 Architectural blending with coving, expands viewsheds and space.

,Learning by Experience

While coving was quickly embraced, BayHomes had an evolutionary path that consumed almost a decade to perfect. BayHomes interior space as a major component of the overall neighborhood design meant that architecture needed to be considered as a primary element at the initial concept plan stage. Initially many, but not all, developers and builders welcomed Architectural Blending into their marketing arsenal. Today, most realize the additional value to attract home buyers and renters. Home buyers need to perceive they are getting greater value to invest in something better.

In figure 10-17 shows viewsheds from within 'living spaces' of homes within Greenbriar Hills, in Buffalo, Minnesota. With architectural blending, viewsheds ('arrows' on the plan), become reality - *but only when the builder is aware of these advantages*.

A New Era of Communication and Collaboration

We also discovered that architects embraced planning methods which elevate Architecture's importance in the land development design process. The collaboration takes on a new dynamic not previously experienced in the suburban land development arena. The goal is to create a neighborhood that enhances living quality, while contributing minimal (if any) construction costs.

Thus, architecture in Prefurbia becomes a critical component in elevating livability.

The rendering below (Fig. 10.18) represents a collaboration between Jamil Ford of Mobilize Design, a Minneapolis architectural design firm, and our planning firm. These affordable townhomes are designed to provide a panoramic view of the surrounding open spaces on every unit for the low income families that would reside within the neighborhood.

The builder recognized the obvious advantages of 'blending' interior spaces with exterior spaces in concert, representing an entirely new way to provide affordable, yet desirable housing. Architectural blending as well as most other new methods is taught more extensively and with greater detail in the LandMentor system.

Figure 10.18 Affordable townhomes with panoramic views - architecture & rendering by Jamil Ford of Mobilize Design

Landscape Architecture

Why create viewsheds from within a home, if the landscape architect designs a theme with shubbery that blocks the views? The Landscape Architect is not only in charge of creating an attractive theme, but is also the gate-keeper of protecting viewsheds. In Prefurbia, the landscaping must preserve and enhance views as well as help enhance the preception of space. Thus, Landscape Architects are key members of the overall collaborative design team.

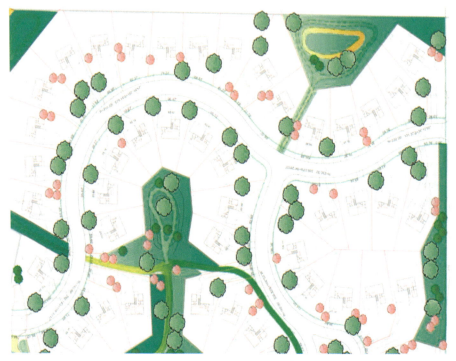

Figure 10.19 Landscaping from a birds-eye view is seen much different than on the ground view. A great Landscape Architect will honor the views. Example shown is Greenbriar Hills by Paul Miller Design.

In Fig 10.19 the landscape theme must be in harmony with interior spaces and the goals of the neighborhood plan - the gatekeeper of views that are maintained as materials mature.

Another caution is to always make sure the landscape architect designs for full maturity of materials, not just for the initial sale. We have observed, over the five decades I've been in the design business, many developments become overgrown with excessive landscape materials that should be removed to enhance neighborhood character.

Reducing Architectural Waste

In this book we introduced the importance of recognizing site design waste and introduced value enhancing methods, such as utilizing 'scale' to increase the perception of space. The same practices apply to architectural floor plans. As stated before, a great architect can make 2,000 square feet have the feel of a 2,500 square foot space. For example, if a 2,000 sq.ft. home has 40 feet in total hallway space, that is 40 X 4.66 (4 foot wide hall + wall thickness), or about 10% of the home interior as waste. This waste could have been used for active areas (kitchens, bathrooms, etc). Too often we see homes designed too quickly, as with site plans, as if there was a competition or race to get the job out the door, forever sacrificing efficiency and value.

To demonstrate the difference in the feel of space that panoramic views provide, look at Fig. 10.20 representing a typical suburban home or townhome with a kitchen overlooking a dining or living area. The area is 'closed in' with a nice view out the rear but only 'if' you are standing at just the correct location within the living room. Limiting glass also limits focal points where quality views can originate.

Fig. 10.21 presents a better living environment and more value. From everywhere on the floor there are great views. This same technique can be applied to all forms of housing, even urban high rise towers, where floor plans can be staggered enough to allow a window wrap. In a grid, with every home in a row, this advantage vanishes.

Staggering is an example where a 'forms based' (Smart Growth) ordinance falls apart, as it suggests only rigid relationships.

As far as reducing waste, to produce the types of views you see here means building fewer walls, and more windows.

Figure 10.20

Figure 10.21

Less materials equates to less cost. By reducing development construction costs, more funds become available which could be used to create nicer architecture and landscaping. By taking this additional step, and reworking interior space to eliminate waste, it might release enough funds in construction efficiency alone, to pay for the extra glass.

Viewsheds Through Interior Spaces

Superman may see through walls, but nobody else can. Reducing the average square footage of a home reduces energy consumption. Instead of convincing the home buyer to desire a smaller space, why not make a smaller space feel satisfyingly large? Space can be 'expanded' in a variety of ways.

For this next example, we designed a floor plan with a panoramic view from the front corner of the home which overlooks a wooded area from the main first floor greatroom. The master bedroom which is on the second floor adjacent to the greatroom is where these pictures were taken.

Looking at Fig. 10.22, you can obviously see how much 'value' *no view* adds to this upper floor bedroom.

By using an interior window (in this case, two of them) views can expand through the home to further take advantage the blending of interior and exterior spaces (Fig. 10.23).
There are a variety of ways interior space can flow from room to room. Openings with closeable shutters work well and cost less, but they do not deaden sound like a window does. This technique in space expansion cost but a small fraction of the enhanced value it creates.

Figure 10.22

Views Through Exterior Space

To create a sense of space, it is important to understand what are the qualities of the space that are to be overlooked. For example, overlooking 5 acres of pond adjacent to a home provides a much different feel of space than overlooking 5 acres of heavily wooded land with thick underbrush. The planner must not just look at a land survey drawing, because 'open space' may not actually be 'open'.

In northern climates after leaves fall, open space will tend to expand through trees, but in southern climates there may be no fall season.

Architectural Blending merges land planning and architecture and as such takes on a far different mindset than subdivision platting. In order to create better value and living environments we need to change industry-wide behaviors. This means we need to redefine how the entire land development industry operates, communicates, and collaborates.

Figure 10.23

Placing a Square Block in a Triangular Hole

Since the birth of the American suburbs more than a half century ago, planning has somewhat moved away from the rigid grid. Since the 1960's suburban design embraced curved streets and cul-de-sacs. A curved street introduces two new lot shapes: the 'inside' and 'outside' of curve 'pie shaped' lots.

Before suburbia, the world's landscape was primarily designed with rectangular lots and rectangular homes. Today, millions of suburban lots have been designed more 'pie shaped'.

These pie shaped lots (along curved streets) provide architectural opportunity - yet, builders still place rectangles (home pads) within them. How does a rectangle to fit into a triangle? Not very well, as the rectangle must either be made smaller, or the triangle bigger - both equate to waste.

Prefurbia begs the question: does the rectangle become reduced to fit in the triangle, or does the triangle become bigger?

Why not make a home 'shape' something other than a pure rectangular 'pad'! Land Surveyors and Civil Engineers do not see the home as anything other than a rectangular 'pad' for their calculations, and most other planners see the home envelope as somewhat of a square.

We previously introduced the use of 'scale'. For coving, this means an aggressive change in home formation along the streetscape to increase the feel of space. Coved lots are pie shaped.

To increase *perceived space* when inside a home, only slight changes, even as small as a foot, can make a large difference This can be felt when walking inside a home under construction as it is being framed, before the walls are covered. Standing in the home and in the areas such as a bedroom or bath it seems that these areas are very small, because there is an awareness of greater space all around. Once the walls are completed and rooms are finished they seem larger because the scale one senses, changes. When viewing from a street, small changes in scale cannot be noticed, it takes an aggressive change in distances to create an impression of expanded spaces.

Architectural shaping becomes more important as density increases. Urban lots have evolved from being narrow but deep a century ago, to the 1960's, where homes were wide but not so deep. In the more recent pursuit of density, land development has been evolving back to a narrow lot. The 'narrowing of homes' if for economic (affordability or greed), or anti-sprawl reasons results in living spaces being more compressed, making it ever more important to make every square foot count. BayHomes solve this problem, but not everyone will desire them, as they are intended for a specific type of home buyer. The problem with a narrow home on a traditional single family lot is how to fit sufficient garage spaces without creating a mundane 'garage-grove' streetscape.

The answer is in Prefurbia: *Architectural Shaping.*

Architectural Shaping

At first glance the single family site plan (Fig. 10.24) appears similar to others in this book. Note how coving allows inside pie shaped lots to be abundant on the outer part of a curve creating more inside pie than outer pie lots on this section of 45' wide lots in Trasona at Viera, Florida.

By expanding width of the front of the 35' wide homes, that would have otherwise been dominated by garage, they are now dominated by the front porch. This makes the facade of the home

Figure 10.24 Trasona of Viera, FL, both front and rear of the 'pad' are shaped to the lot. Homes designed by Annette Turner www.UbiquitousDesignsInc.com.

appear as if on a 60' wide lot! The outer pie lot homes widen towards the rear to 41' allowing for better interior floor planning and greater ties to the larger than typical rear yard spaces. As seen on Figure 10.24, the master planned community takes on a whole new meaning to livability, function, value, and beauty. In the case of this example, the home became larger than if it were to fit a rectangular 'pad' size, however the homes could have maintained a target square footage without excessive depth increasing rear yard space. Note on Figure 10.24 how most lot lines parallel a home side to help residents easily locate their side lot line, a method often used in coving.

Annette Turner, a home designer for Viera Homes, was furnished a set of dimensional targets to strive for in her floor planning as well as major view criteria. She developed a series of home plans that embraced both Architectural Shaping and Blending. After architecture was determined, we then created the lotting of the site plan. While this approach is not as simple as 'just ignoring the home', it eliminates as much waste in design as possible. Most important, it provides curb appeal that no other development at the same density in the region can compete with! Results? In the first year of sales they sold almost 200 homes! Other builders that used shaping have reported similar success.

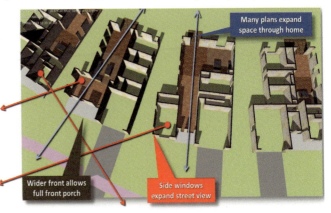

Figure 10.25 Home/Lot space & view integration requires collaboration of planner, architect, builder, developer, and engineer to create this neighborhood.

What about Existing Regulations?

Architectural Shaping as well as Architectural Blending does not break any regulatory minimums. In fact, there are often loopholes in regulations that allow a significant increase in density (up to 20%) when Architectural Shaping is used along inside lots on curved street subdivision or coved designs (again, without reducing a single minimum).

Figure 10.26 Trasona aerial and location on original plan - these homes are the narrow 35' wide product - shaped!

Architectural Shaping provides a significant market edge for home builders to gain leverage over their competitors. When combined with Architectural Blending, it creates a greater advantage. More important, it creates more long term value, or in other words - *sustainability*.

As mentioned many times the methods and techniques introduced in this book require a major shift in behavior for the consulting industry from 'producing quick plans', to attention to detail demaning greater effort while also investing in the proper technology and training. Trasona at Viera as made possible by the civil engineer, Hassan Kammal demanding that the developer use our methods! Hassan of BSE Engineering in Melbourne represents a civil engineer who cares more about the neighborhoods they create than getting out a quick and dirty plan.

None of this new era is possible without dedicated civil engineers, architects, land surveyors (all who act as Land Planners), as well as the willingness of land developers and home builders to provide the best possible product to their customers. The by-product of which serves the needs and demands of the municipality with sustainable growth.

The how-to's of Architectural Shaping are far more cpmprehensive than can be addressed in this book, and require a higher level of technology than current CAD based systems and their training.

For those that argue against this new era of design, you must ask why - is it because they will not make the time to provide the best for their clients?

CHAPTER ELEVEN

Prefurbia Methods for Urban Re-development

In the late 1990's we researched how Prefurbia methods could addresss urban redevelopment. At that time the cost of land (to relocate existing residents and tear down their property) per acre compared to the suburbs seemed impractical. The economics did not make sense . For example, (at that time) a suburban Minneapolis acre with city services sold for about $30,000. The same time in 'the city', assuming an existing home density of 5 homes per acre, to remove homes would mean having to reimburse the 5 families at a total of about $1,250,000 an acre because there are few 'downtrodden' areas in Minneapolis. Replacing an acre of decent housing just to build a new mixed use development, requires large subsidies and extreme density to make any financial sense.

However, lower value areas such as in Detroit or East St. Louis, is feasible to tear down blocks of land and start over. We created an Urban Prefurbia model for a HUD study (below), to demonstrate the advantages of Urban Prefurbia. An existing 10 acre sample from a GIS map data of the City of South St. Paul was chosen for the study, as shown in Fig. 10-27. There is a school to the west and a main street to its east.

The larger rectangle on the right is an apartment building and the remaining pink rectangular shapes are single family homes that have one and two car detached garages.

For analysis we used LandMentor which easily reports precise areas of structures and paved surfaces. It is important to have accurate data when comparing advantages of new designs.

A chart (see Fig. 10.28) shows that each unit averages 5,272 sq.ft. of 'man made' improvements (buildings and pavement).

The average home footprint consumes only 1,372 square feet of the 5,272 square foot total. Thus, there are 3.8 times the square footage in streets, alleys, garages, and walks as there are in the footprints of homes. This high ratio of floor area to paved surfaces is common in urban settings.

The unit breakdown and density is shown on the chart in Fig. 10.29.

Figure 10.27 A 10 acre example site as a basis for a HUD study

We used this example because it is typical of residential development just outside an urban downtown core.

The Prefurbia model abandons existing internal right-of-way, creating contiguous areas, while retaining usable utilities under the abandoned streets. Just as in suburban design, we create the pedestrian systems first.

An 8 foot wide main trail was placed cutting a straight tree-lined path diagonally through the site as seen in Fig. 10.30. This trail also serves as an emergency vehicle access.

The north, west and south boundaries feature single family coved lots with homes that are shaped (Architectural Shaping) to take advantage of the lot configuration.

The coordination between interior spaces and surrounding open areas (Architectural Blending) is seen in Figure 10.31 for both single family and multi-family living.

This particular study (we did several) kept the single family appearance for neighbors, while delivering a sense of space that never existed before in this residential area close to the urban core.

While some of our studies demonstrated a wider variety of home types and housing price points, in this study we wanted to provide affordable apartment living with panoramic views of open spaces (seen in Fig. 10-32) as well as single family homes within similar price points of the region.

Average Impervious Area Per Unit

Public Paved Surface Area:	1,913 sq.ft. / unit
Private driveways & alley (28 Homes)	886 sq.ft. avg / home
Private driveways from street (16 Homes)	856 sq.ft. / home
Sidewalks, walkways	431 sq.ft. / home
Garages	429 sq.ft. /home
Single Family House Footprint	1,372 sq.ft.
Total AVG Impervious Surface Area / unit:	**5,272 sq.ft.**

Figure 10.28 Average Impervious surface area per unit of the existing conditions.

'As-is' Site Information

Site Area:	10.8671 Acres
YIELD	Totals
Total Single Family Homes	49
Total Apartments	8
Total Units	**57**
Total Garage Spaces	80
Density: Gross	5.25/Homes per Acre
Net	7.46/Homes per Acre

Figure 10.29 Current density breakdown of the existing conditions.

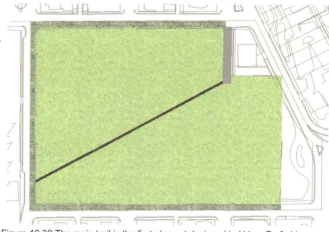

Figure 10.30 The main trail is the first element designed in Urban Prefurbia.

For traffic passing by the development, it will appear as it contains only single family homes.

LandMentor computes and reports area of improvements. This 'area' determines many things: storm water run-off, efficiency of a particular design, environmental impact, etc.

Critical data can be charted in a variety of ways. The default we coined

Figure 10.31 Blending interior living spaces and views with surrounding open space.

Environmental Density, which reports total volume of man made surfaces, thus allowing a proposals impact to be measured. Without this new technology it was not possible to easily know how efficient a land plan was - *at initial design stages*. It was the ability to develop spatial software in the early 1980's that made Prefurbia possible.

With 'Urban Prefurbia', tracking impact is just as important as in suburban planning.

In the developments surrounding the urban core, the Prefurbia method begins with the removal of excessive infrastructure and abandoning street right-of-way, then apply the more efficient designs introduced in this book.

Figure 10.32 Maintaining panoramic views within every apartment unit adds value.

It would not be unusual to see a 50% or greater reduction of infrastructure while also increasing density compared to an existing urban grid!

Chapter Eleven: Urban application of Prefurbia   149

...This study included two asymetric three story condomimium buildings with underground parking (2 cars per unit).

The results were outstanding, not just in the beauty of the neighborhood, but in the 'numbers' compared to the before (existing) plan.

Prefurbia increased density 25% from 57 to 76 units, however, more important; the pavement area plummets and available covered parking is greater while monotony is eradicated. *Environmental Density decreases to just 35%!* The site plan provides panoramic viewsheds from every home and utilizes low impact landscape design methods. Again, the 'tree-lined' 8' wide trail bisects the neighborhood providing alternate emergency vehicle access. This creates a wonderful place, a preferred setting in which to live.

   Another critical issue solved is parked cars. A large architectural and landscape budget may create character, but as soon as unsightly old cars are parked along the streets, there goes the neighborhood! 'Car-scape' is a major concern in low income housing developments.

Prefurbia softens the impact of vehicular clutter as can be seen in some of the above images derived from LandMentor's Virtual Reality.

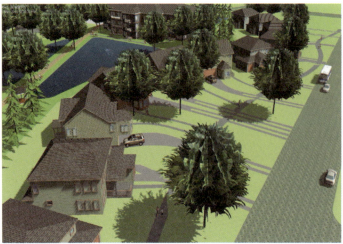

This (Fig.10.33) image shows how the increased setbacks of coving (when applied to rectangular urban blocks) increases space along straight streets.

Note how much distance there is between the homes and the curb line creating 'parks' along the front yard while also providing park spaces in rear yards. Porches dominate the streetscape yet homes are 'garage forward'.

Figure 10.33   The large coved front yard provides openness that is competitive with any suburban development utilizing existing urban streets.

Low Impact Landscaping

Common areas that are association maintained can utilize low impact landscaping such as this 'no mow' fescue (Fig. 10.34). This is a viable sod alternative.   Advanced landscaping techniques reduce the cost of maintaining the added open spaces that Prefurbia provides.

The yard in Fig. 10.34 is the author's. It was only mowed twice in the first five years. However, a tree fell during a storm that allowed weeds to take over. Again, caution must be used when adopting new age landscape alternatives. We learned no mow as well as other so called 'maintence free' options are not actually maintenance free. Caution should be taken that rain gardens, manmade prairies, no mow, etc. all need continual maintenance or they quickly fail. When properly maintained, they should still represent a savings compared to sod - maybe.

Figure 10.34   No mow and no watering yards are not science fiction - they are a fact. The lawn is more sculpted than sod, but acceptable as a replacement in suburban settings whare manicured space is important.

Another new innovative solution is from Irrigreen LLC, who developed sprinkler systems that can water landscaped areas in any shape, thus reducing water useage in half! More information on this exciting new product can be found at: www.irrigreenllc.com

High Density in Prefurbia

This image (Fig 10-35) is an example of a highrise density blended with suburban residential settings. The only caution should be to locate elevated structures in places where shadows do not interefere with low rise residential and quality views surround tower units. Note LandMentor also provides accurate shadow mapping.

The Cyber Village

This concept introduced by George Van Hoosen, of Global Green Building, offered a way to serve the 30% of homes that operate a business out of them.

Fig. 10-36 shows a study prepared for HUD using the 'Cyber Village' model. Each of the larger homes have an office for a small home business to operate. The home 'office' area is shown in purple below. To supplement and support the 'home business', a central Cyber-office (orange) with mail room, receptionist, meeting facilities, and larger offices can be contracted for use to the residents.

Figure 10.35   Prefurbia blending urban high density towers with suburban settings.

The Cyber Village concept is an idea that makes sense in today's connected world. An expansion of this idea would include nursing and medical facilities.

It can be utilized in urban and suburban developments, no matter what planning style the developer chooses.

For more information on the Cyber Village concept see:
www.globalgreenbuilding.com

The Chandler in Frankfort, Kentucky has five of it's 90 units as home businesses as part of the PUD.

Figure 10.36   The Cyber Village Community/Office Center.

CHAPTER TWELVE
A Model Coving Ordinance

"A common mistake that people make when trying to design something completely foolproof is to underestimate the ingenuity of complete fools."

— Douglas Adams

No matter how good of an idea, concept, or method there will always by ways to make it go terribly wrong. Take for example, New Urbanism, a simple design concept very well explained (cnu.org), yet when its basic elements are misunderstood and altered, it can fall short or fail.

In the New Urbanism, use of cars is discouraged along with the visual impact of garage doors. Alleys are the norm. Grand porches connect to walks in front of the home, to strengthen community interaction. Common public areas throughout the neighborhood enable resident interaction. New Urbanism encourages employment, entertainment and shopping within a short walk from home. So simple, yet it gets corrupted upon implementation on a regular basis. The same is true of Prefurbia - we have seen many instances where developers tout coved developments that have been terribly designed, inefficient, or simply - not coved.

Any planning method, even if everything goes well during the approval stages, after approval, the city may have just a recorded plat with nothing in place to guarantee the critical design elements will actually get built as promised. Lots are sold and home builders may not be aware of strict adherence for landscape and architectural details. If elements 'after-approval' are not held, a formula for failure exists. Again, this is both a New Urban and a Prefurbia problem.

A simple grid plat is needed for New Urbanism. Coving requires complex 'plat' geometry. Unlike New Urbanism, coving does not specify architectural or landscaping detail. However, in Prefurbia, coving does. Coving improves curb appeal by adding shape, space, and beauty, independent of architecture and landscape, thus enhanced marketability. Character-enhancing elements of architecture and landscaping in the developments our firm designs, are most often mandated. Walks, pedestrian connectivity and parks are not a 'requirement' of coved design. Those traits are demanded in Prefurbia, not coving.

Regulations can be written to encourage character-enhancing elements independent of any planning design style. This chapter will show how to create a 'rewards-based' ordinace using wording for a 'model coved ordinance' as an example Thus, the model ordinance contained in this chapter can be used as a general guide to revise regulations that will create more sustainable growth by rewarding developers for exceeding the minimums.

We worked closely with many other consultants to learn and experience how to best serve the client, city, as well as the developers and builders that contract our services. We develop a system that cities can use to make sure what is presented will be built as intended. Some of these safeguards would also assure that New Urbanism is built as prescribed.

With the model ordinance that is contained in this chapter, we create a 'tool set' enabling cities to motivate land developers to exceed the expectation of a minimums based system. Staff can have a simplistic set of methods without the cumbersome process of approving variances or exceptions.

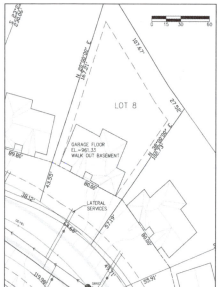

Figure 12.1 A recorded coved lot detail

Coving from an engineer's perspective

Implementing the design of coved neighborhoods into platted subdivisions can be a challenge in terms of translating the ideas of the land planner into stamped documents and recorded plats.

Engineers are conservative and linear thinkers. The engineer's motto "that is the way we have always done it" is no excuse to halt progress.

To get the team on board with 'outside the box' thinking is a challenge. The charge of an engineer and surveyor is simple: take the land planner's concept and make it a reality. We have found too often, it is easier for engineers to fabricate as many reasons as they can, to fight fresh ideas, than to work towards a making better neighborhoods (Figure 12.1).

Of course, this isn't always the case, but a gentle reminder that the customer (either the developer or ultimately the home buyer) comes first, often helps.

Coving's varied and meandering setbacks add an extra layer of dimensional control (effort). Meandering walks also require that easements be defined (more effort).

As professionals find issues that need to be addressed (adjusting lot lines for utilities, as an example) they must communicate with the land planner to make sure that any required revisions do not compromise the integrity of the plan. The land developer should designate a single point of contact among the surveyors and engineers, to discuss plan revisions required to maintain the planner (and developer who hired them) vision. Building collaboration and communication is key to both Prefurbia and New Urban successful neighborhoods.

Figure 12.2 Yikes! Coving is NOT an extreme stagger of homes as illustrated here, a problem created by the engineer.

Example: holding to the setbacks

Coving sculpts the streetscape. Coving does NOT stagger the homes (as shown in Figure 12.2) - they 'meander'. In the above case, the civil engineer destroyed the design setting homes in a haphazard pattern, made worse with home fronts looking into home rears!

Settler's Glenn is a wonderful 'first generation' coved neighborhood in Stillwater, Minnesota. However, there are two homes that the engineer changes without consulting us which detracts from the neighborhood character, ever so slightly, but it decreased curb appeal - thus, lowers value.

The original plan (Figure 12.3) shows the home fronts all facing the main street and the entrance to the cul-de-sac, lots 11 and 24, having the homes angled into the corners. The engineer changed orientation of homes at the intersection. As you can see (Figure 12.4), the side of the home on lot 24 lacks any architectural detail and windows. Worse, residents and visitors, as well as the home across the street, directly view into rear yards. While this is a minor infraction, it would have been solved if close communication been established between the engineer and planner.

Figure 12.3 The original approved Settlers Glenn site plan

Figure 12.4 The Civil Engineer destroyed the nice home setting

Additionally, if coved regulations conflict with existing engineering standards, then municipal engineering and emergency services staff may be reluctant to place their stamp of approval. If there are conflicts, then changes in those standards should accompany the new coved ordinance so as not to adversely impact the efficiencies achieved with coving. For example, there is no reason sewer manholes need to be on the centerline of a street other than *"that's the way we always did it"*

Consulting engineers are key to convincing their municipal counterparts that changing strategies does not compromise the integrity of the infrastructure. Typically, the municipal staff will not believe the land planner and will only see any 'relaxing' of standards as a ploy by the developer to save money. Forming a level of trust with staff will make the approval process much easier when they recommend approvals to the commissions and councils.

Once plans are approved and lot sales start picking up, it is time to make sure that the coved neighborhood comes to life as represented (again, this also applies to New Urbanism). In many instances the plats are recorded and filed away, but the *building inspectors* might never been informed of the unique setback requirements on individual lots needed to implement coving.

Typically, builders and home owners come in with their plans to the building department to pay their fees and get permits, without being told that building restrictions apply to the individual lots.

We have found it very useful to simply create a binder for the direct use by the building inspector on each individual lot. Each lot is shown on a single sheet of paper, to scale, that is an enlargement of the final recorded plat. The diagram shows all the pertinent easement information, building envelope size and location, as well as the recorded setback for the lot. The binder becomes a working document organized by block and lot within the subdivision, so that all parties can understand the nuances of the lot they are building on. It is kept on file with the building permit department. We also suggest that real estate teams and builders working in the subdivision have copies, so that they can share information early on with prospective buyers.

Critical items

The Titanic sunk because of a lack of communication. Many lives could have been saved if communication with a nearby ship had not been misinterpreted. The same holds true in any new development that does not follow the standard cookie-cutter approach. Just because a development sells fast and is profitable, does not mean it offers better quality of life. For all the features, functions, and benefits we explain to the city staff, to receive approvals on these neighborhoods, it is astonishing that when we visit the developments, the builders own sales staff seem to be ignorant of the many neighborhood benefits and instead touts only kitchen and bathroom!

Too many developers go into 'sell lots to builders' mode, assuming that the builders will know the benefits and then communicate them to the sales manager, which will communicate all of those benefits to the sales staff. If the potential home buyer walks away (especially in the early stages of development) it is often because the neighborhood advantages were not explained. This is the most egregious marketing error that occurs. Today, LandMentor supports for the Microsoft Mixed Reality headsets to deliver neighborhood plans in a Virtual Reality format. To help home buyers understand what the completed neighborhood should look and feel like, we typically deliver the neighborhood digitally in a VR format along with the 'free' LandMentor Viewer. This allows anyone to experience the actual 'feel and space' of the development as if 'being there'. In fact, we were the worlds first consulting firm to use VR during site plan approval (July 14th, 2015) in Springfield, Nebraska for Graves Development Resources. The team at Microsoft was instrumental in our developing the technology that is now part of every LandMentor system.

Another marketing mistake is selling random lots. Unless VR is implemented at the sales office, the consumer cannot perceive the 'scale' of coved setbacks, - until a consecutive series of homes is built. To randomly sell lots and wait for the shaping to occur is a too common and costly mistake. We have noticed that developments that initiate building along a continuous deeply-coved setback, sell homes faster than when homes are constructed sporadically! Again, implementing a LandMentor Viewer VR system at the sales office helps.

Justification for a model rewards based ordinance

In Chapter 3, we shared the tale of two developers, one that stood out as a developer who did everything to the minimal requirements and the other who went out of his way to do the best for the community. This is because they were both using the same *minimums-based* set of rules. If there were a *rewards-based* ordinance – the reward being 'density', it would lead all developers (even the developer who cut corners) to strive for a higher standard, thus achieving sustainability.

We were the first planning firm to use VR at city meetings for approval - above is Springfield Pines in Nebraska

Other than coving, all of the techniques discussed in this book can be applied to any form of planning. The concept of 'flow' can be applied to any development, the idea of creating walks that lead *through* a neighborhood, separate of the street system, is not limited to a planning style. Zoning transitions can be reversed, and used in conjunction, or combined with streets that flow and convenient walking connectivity.

The following model ordinace rewards those developing more sustainable neighborhoods instead of constructing subdivisions which would otherwise result from adherance to minimums.

A MODEL REWARDS BASED ORDINANCE

This ordinance presents a design standard for single-family coved neighborhoods using a rewards based mechanism added to a set of minimum standards.

Note: Numeric figures that are variable will be designated AA, BB, etc.

General

Proposals for coved neighborhood plans are subject to all procedures and specifications in the existing subdivision regulations, except as noted herein.

Blocks

With a free-form design such as coving, it is difficult to determine a block by length. Instead 'blocks' will be defined by perimeter length of street and walk right of way.

Minimum block perimeter length is not needed, but the maximum shall generally be 2,500 feet as measured along the right of way as street or through walk. The block perimeter length can be greater or less where topography or configuration will not allow conformance (Fig 12.5).

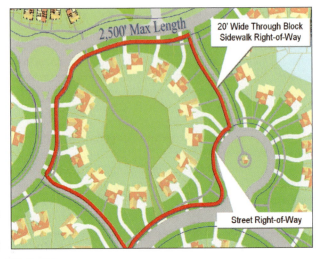

Figure 12.5

Lots

A 'cove', in planning, is an indentation of the front setback line intended to: a) create a park-like (spacious) streetscape, b) reduce road length, and c) create 'scale' from street view, encouraging angles of views from within homes..

A coved lot is similar to a conventional lot, except the front setback along the lot line is individually dimensioned when the distance from right of way is greater than the minimum front setback, as defined by conventional subdivision front yard minimum.

Staggering of the front setbacks shall be avoided (but not excluded), instead using a meandering setback designed to open up views along the street, avoiding monotony.

A proper coved design uses the home fronts (the setback line) to form a curve that differs from the curvature of the street (Figure 12.6):

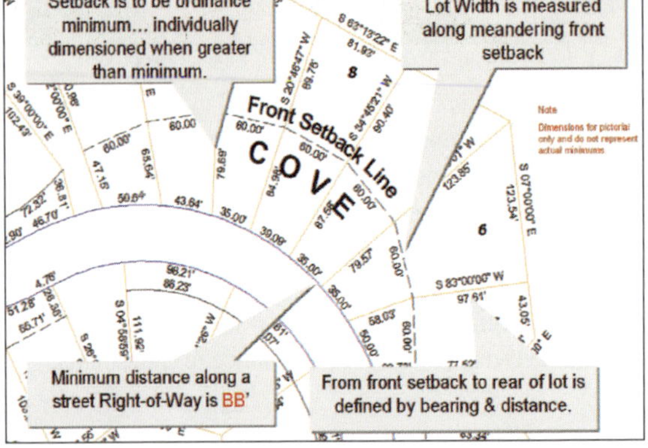

Figure 12.6

- The minimum lot width along the meandering front setback line shall be AA feet wide, as measured from a point where the front setback line intersects the side lot lines.
- The minimum lot width along a street right of way for *frontage,* shall be BB feet.
- The layout should avoid angles on side lot lines unless it is difficult to meet the BB feet right of way length minimum.
- Each side lot line shall contain a dimension distance from the right of way to the front setback intersection, then a bearing and distance from the setback intersection to the rear lot line shall be on the preliminary and final plat. An additional bearing shall be shown if there is an angle point on the side lot line, otherwise it is assumed the side lot line is void of angles.

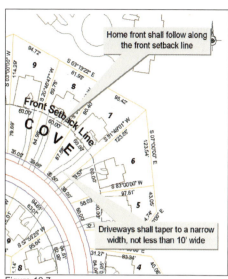

Figure 12.7

The home front shall be constructed at the same angle as defined by the front setback line, i.e. perpendicular and parallel to the front setback line.

Side lot lines need not be perpendicular or radial to the right of way line (Figure 12.7).

Longer driveways (those that are longer than 30 feet, garage door to curb line) shall taper to a narrow width at the curb (determined by the vity or the builders desired minimum).

Side yards

Side yard minimum shall be 5-feet from the side lot line to the home structure.

Overall *average* side yards on coved neighborhoods are thypcally greater than conventional platting, as few home sides parallel each other.

Side yards on a corner lot shall be 10 feet from the right of way line.

Street pattern

A coved street pattern shall be designed to reduce the number of streets, using a meandering pattern, yielding a reduction of intersections, compared to conventional parallel streets. Streets shall be designed to reduce speed by avoiding long straight street patterns and also to avoid unnecessary obstacles to traffic flow (Figure 12.8).

Figure 12.8

If a street splits to form a landscaped island, a variable meandering (non-symmetric) pattern is encouraged. The right of way shall parallel the meandering pavement.

The one-way lanes shall be a minimum 20-feet wide.

A minimum landscaped island width of 12 feet shall be maintained.

Tangents

No tangents are required nor are they encouraged.

Streets and right of ways

Street construction shall conform to existing subdivision regulations

Coved designs encourage much larger radii for cul-de-sac , to increase efficiency. One way lanes are to be used with a center island on all cul-de-sacs.

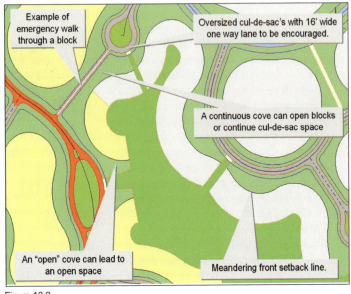

Figure 12.9

Coved patterns

Gentle and gradual transitions from minimum setback to deep setbacks should be avoided, as this will dilute the sense of scale. A more aggressive, shallow to deep transition should be encouraged (Figure 12.9).

There are many different types of coved configurations. In general, the coved indent pattern and setbacks should vary in depth and shape throughout the neighborhood to reduce monotony.

Lot Area		
	With Walk System	Without Walks
Minimum Lot Area	C,CCC sq.ft. (smaller value)	E,EEE sq.ft. (larger value)
Minimum Average Lot Area	D,DDD sq.ft. (smaller value)	FF,FFF sq.ft. (larger value)

Lot size

To encourage pedestrian-oriented coved neighborhoods, there shall be two lot size standards. If the proposal does not create a dedicated pedestrian system (see below) the larger lot size will apply.

Lot width

Minimum lot width shall be measured along the meandering front setback line. Minimum front setback from right of way measured along lot side line or parallel to the right-of-way shall conform to existing subdivision regulations.

There is no maximum lot depth, however, rear yards ending at a point should be avoided.

Rear yard

Minimums shall conform to existing subdivision regulations.

Green space

Lot Width		
	With Walk System	Without Walks
Minimum Width	AA feet (smaller value)	GG feet (larger value)
Minimum Width Along ROW	BB feet (smaller value)	HH feet (larger value)

Any park, open space and/or green space standards shall conform to conventional subdivision standards.

Pedestrian-Oriented design

Walks shall be designed to allow the shortest distance through the neighborhood, regardless of the street pattern. The intent is to make it as convenient and safe as possible to walk through the neighborhood.

The following design principles shall be encouraged:

Meandering Walks

A system of meandering pedestrian walkways is more attractive, and safer, by separating pedestrians and vehicular traffic as much as possible. A meandering walk can be within the right of way of a street but also outside the street right of way. When the walk meanders outside the street right of way, a public easement shall be dedicated and defined to be parallel to the walk and 2-feet outside the walk's edge. This walk easement shall be designated on the preliminary and final plat.

Width

Where walks through blocks also serve as an emergency vehicle path, the minimum width shall be 8-feet wide in a 15-feet wide walk right of way.

Walks on one side of the street shall be preferable to walks on both sides of a street, except on high traffic volume streets. Walk width, when on a single side of the street, shall be 6-feet minimum, and 5-feet minimum if on both sides of a street.

Proximity to a home

A meandering walk shall not be any closer to the front or side of a home than if that same walk was constructed using conventional design standards.

Rate of Meandering

Walks should be gently meandering using large radii, avoiding zigzag patterns.

Additional right of way and street design parameters

Right of way width along one-way street sections shall be 40-feet wide.

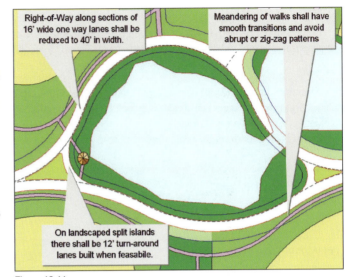

Figure 12.11

When width of an island allows, there shall be a 10 to 12 -feet wide turn around lane provided in islands, shown in Figure 12.11.

The above model ordinance addresses standards that should lead to sustainable coved neighborhoods. However, as you may have noticed, density was rewarded by adding the pedestrian system - a 'reward based' method that could easily expand to better architecture and landscaping or for that matter, anything the city desires! The minimums stated here are determined by standards passed by the many cities we have submitted and are reasonable. Some of the city staff may ask for other standards, and that's typically fine. For example, the walk used as an emergency vehicle alternate route being 8' wide is what most cities have approved. Some might want wider widths, which is fine - the danger being a walk that can also seem like a street encouraging local traffic shortcuts. The dimensions we have listed in this chapter are based on the 26+ years experience with the design methods used on over 1,100 neighborhoods we have been contracted to

CHAPTER THIRTEEN
Technology & Education

"Education is what remains after one has forgotten everything he learned in school."

— Albert Einstein

Four decades ago Rick Harrison created 'Hewlett Packard -80 Series CivilSoft', his first 'site design' software product. This $2,000 package would perform land surveying and civil engineering calculations and accurately plot to scale using thermal paper. The software distributed through a local (Dallas, Texas) land surveying supply dealer. There were no support calls, ever. It stood to reason, no support calls meant nobody was using it. The same time, he came to the realization that no matter how great he thought his planning skills were, his land plans ultimately became bland cookie-cutter subdivisions.

His plan was to abandon both the planning and software business and find a new career path.

The future was changed by a single phone call...

The call was from a Vice President at Hewlett Packard, who investigated all of the engineering software products advertised on the market and found the HP-80 CivilSoft to be on top of the 50 surveyed for customer satisfaction!

That was why there were no support calls!

The next day Hewlett Packard shipped the first prototype HP-87 computer (the most powerful desktop computer at the time) to Rick. On the day the HP-87 was introduced, so was his *HP-87 Site Computation & Design*. It was the first of many successful land development software packages, and where today's 'LandMentor System' has its roots.

Rick Harrison programming the HP-85

The software business was tough to keep up because of constant hardware and operating system improvements, but massively profitable. That's Rick below, flying his company's Piper Malibu in 1987, on the way to sales or software training. It was during the 4,500 hours in the air, where Rick looked down on the geometric blandness of the nations new growth. His own software and his competitors had influenced mundane design by automation - instead of thoughtful skilled design.

Technology was actually making growth worse!

Rick used the knowledge gained in the software company to research new 'design' methods. In 1990 he began the process of harnessing technology to create better plans. This was the start of the collection of methods collectively called Prefurbia.

Technology cannot provide sustainability

Pictured below is a surveyor's 'chain' - a device that has not been for well over a century. It is 66 feet long - made up of 100 'links'. Many of today's street right-of-ways are 66 feet wide. Why? Because well over 100 years ago a surveyor simply stretched the chain across a street to 'stake it out' - *absolutely no other logic!* Today's GPS & Total Stations instantly measure precise locations while storing information that is instantly transferred to the office computer. For the land surveying industry, in the past 4 decades, there's been revolutionary improvement.

For 'land planning', both CAD and GIS technology has no apparent benefit over the past tools - using a slide rule, pen, and paper.

The 'intentional complexity' of CAD surveying and engineering software stagnates progress.

Intentional complexity? A software firms major income is not the initial sale of the product, but the ongoing income from services which includes support, training, and upgrades. Thus, the initial product becomes a 'hook' for future income, assuming the customer is locked into the software for the long term. Updates that change the data structure and subscription schemes force the user into an unending box.

If software could be mastered in a week it would destroy that business model. That was one of our goals to achieve with the LandMentor System we developed.

CAD based software is a terrible 'spatial' platform making it difficult to report impact of the hard surfaces created by land development such as streets, walks, driveways, homes, etc. Spatial reporting at initial design stages is where the spatial information is most needed. Without the ability to report the 'spatial' efficiency of a site design using the Land Innovation technology, of the 1990's, few of the innovations in this book could have been discovered.

A larger problem is that CAD technology automates 'minimums' - *only the minimums*.

Advanced technology + education can foster collaboration

In the four decades of selling software the first question from an architect, planner, civil engineer or surveyor has always been: *How much faster can I get the subdivision out?* The first question have should been: **How much better of a neighborhood can I design using this software?**

A new type of product dedicated to solving industry problems

Existing software solutions rely on a separate CAD or GIS platform. Creating a stand alone 'virtual' precision land development design product from scratch is an expensive and time consuming venture. LandMentor is the first software to introduce a new 'base engine' specific to sustainable land development design. With the help of Keith Willenson, his brilliant partner in software development, they created an entirely new solution. Keith invented a (patented) revolutionary user interface that eliminated the thousands of commands of CAD software, while developing the most powerful precision geometry platform ever created. A goal from the beginning was to harness Video Gaming for interactive 3D virtualization - as a by-product of the normal workflow. Thus, 3D would be easy and free! Working with Microsoft, we were able to create a direct plug & play support for their Mixed Reality technology. This amazing technology implants the viewer into the site for an immersive experience! It will completely change the way city councils and commissions judge a plan.

There would be no 'modules' or 'options' - just one fully integrated base that could foster future innovations for land development. By eliminating months from the users learning process for software use, we free up time to teach how to 'design and think better'.

Thus, LandMentor is as much an enabler, as it is a tool. We wanted to create a new paradigm in the land development industry that would truly make a difference and create a more sustainable world.

In 2018-19, why do most new neighborhoods look as if being designed on an Etch-A-Sketch?

This book is only a beginning...

When we began writing this book, we knew that there must be a mechanism to teach the how-to's, beyond the foundation introduced in this book.

In the early 1990's the profits of the software business funded the research for Prefurbia. In the late 2000's profits from Prefurbia funded LandMentor and the creation of its educational materials that teach beyond the pages of this book.

Keith Willlenson, Rick and Adam Harrison formed the development team. They realized only a solution that blends technology and education can encourage change. Thus, the name LandMentor seemed a fitting product name, which we referred to several times in this book.

LandMentor is intended for everyone involved in the design, construction, and regulation of land development. While still in beta, in 2010, it won the coveted Building Products Magazine Most Valuable Product of the Year Award. In 2011 it won semi-finalist for the TekNE award, and in 2012 semi-finalist for Cleantech Open. In 2013 we received the US patents for LandMentor. It introduces 'positional based' coordinate geometry utilizing precise intelligent spatial data. Software must no longer be about producing site plans using age old methods - faster. Technology must be an enabler to create great sustainable growth.

For more information: www.land-mentor.com

In 1968, if my mother, Carol Minowitz did not get me that interview leading to the land planning job at Don Geake, none of this innovation would exist today. I would like to conclude by thanking my wonderful, understanding, and beautiful wife Adrienne. She has been instrumental in keeping my focus on making this a better world for all to live in. Adrienne is responsible for much of the information within this book.

You can also thank Adrienne for coming up with the name *Prefurbia - a Preferred and Sustainable way to live!*

Afterword

"Sprawl is not evil. In fact, it is good. It is the inevitable result of a free people exercising their cherished, constitutionally protected rights as individuals to pursue their dreams when choosing where to live, where to work, where to educate and where to recreate."

— L. Brooks Patterson, Oakland County Executive, Oakland County, MI.

Every city chooses if they want to embrace growth to create a more vibrant city offering great services - or not. There *will* be growth - some call this sprawl.

Diversity of living styles guarantee there will never be just one planning solution. Standardization prevents cities from growing to its greatest potential. We can no longer be less efficient with land, natural resources, and design methods.

Progress requires change

The education for those designing our neighborhoods must include an understanding of the *financial* aspects of land development including 'market proven' environmental solutions. Plans for growth must have sound economics to avoid a repeat of the recent collapse of the U.S. housing market. If land planning remains something anyone (and everyone) can do, the land development industry won't progress. If nice development only serves the wealthy - 'gentrified' communities, there is no real advancement.

Land planners (design)

Engineers and surveyors acting as land planners who think achieving density by quickly replicating lots and units in a CAD package - is 'land planning', are sorely wrong.

Density does not guarantee profitability. Enhancing living standards and efficiency (both environmental and economic) will make more difference to the bottom line.

The land planner who can create beautifully rendered plans, but does not understand engineering and surveying is obsolete in today's world. Hand drawn plans which did not conform to the reality of engineering had their place in the 1960's, not today.

Those who think 'land planning' can be automatically created by a CAD software are not not just fooling themselves, but also the clients they are being paid to design for, and the cities that they are submitting within. Greater fools are those who develop these automated software packages in the name of good 'planning'.

Consulting planners (regulatory)

Planning consultants who 'boiler plate' (copy) ordinances (based upon 1930's wording) and suggest only minimums, are contributing to the mindless replication of development. A rewards (incentive) layer on top of the existing regulations is what is needed to foster better development for sustainable growth. Ordinances need to be simple and easily understood by all. Progress requires this important change.

Architects

Perfurbia represents a new era of design for architects - blending interior with exterior spaces and custom shaping of buldings to fit lots will raise living standards as well as the importance of good architectural design. This mandates collaboration between architecture, engineering, surveying, and planning. We developed the LandMentor System with an included education to make the integration easier for planners, architects, engineers and surveyors.

Mayor, administrator, planning commission and council

If you do not like the plans, proposals, or layouts submitted for review – and if you find value in what we have presented in this book, demand changes sooner –not later.

Don't chase numbers: Create regulations that 'reward' better development. A *minimums-based* regulatory system will get you exactly what you have experienced: *the minimum effort.*

Make processes as easy as possible for everyone to understand and do not concentrate only on 'numbers'. What difference does it make if you have a 10,000 square foot lot minimum or a 10,250 square foot lot minimum? How does that assure great development? If the planning consultant cannot get beyond a 'numbers-only' focus, find another one. If your planner is more concerned about what the existing neighbors see out their windows, instead of a high quality of living those that will dwell in the new development, it is time to replace your planner.

A great plan presented poorly can influence a "no" vote. Don't be fooled by a great presentation on a poorly planned site. Imagine living in the neighborhood yourself. Imagine residing in all of the proposed homes. Imagine walking through the site. The commercial center design - will it encourage commerce, or in the pursuit of 'social engineering' will it destroy it? Sustainability is about balance. There is no perfect design. If the developer's great design makes them a fortune in profit, that is a good thing. Failed development is unsustainable.

Today, with the technology we developed, all submittals from a proposed shed in a back yard to a master planned community can be presented easily in live action Virtual Reality with reports on the efficiency (and waste) in the proposal making 2D submittals obsolete. No longer does the council, commission, and public need to guess about any proposals.

Educational institutions

Teach your students what they will need to learn to be valuable in the marketplace, not just teaching a preset CAD or GIS class given to your school, intended to make a software company more profitable. Learning how to use the city's' GIS (Geographic Information System) system is worthless as a 'land planning design' tool. Teaching New Urbanism and urban planning is great, but it's not applicable to most of the *suburban buyers* who represent 80% of the housing market. Class projects must promote and teach communication and collaboration. Why not have Engineering, Surveying, Architecture, Planning and Real Estate students all work together on the on the same projects? They will need to do that in the world after they graduate if we ever hope to create a more susainable future. LandMentor is an ideal system for teaching a holistic approach to growth, as it includes a basic education in land surveying, civil engineering, precision mapping, advanced land development deisgn methods, and how architecture should integrate with planning (and much more).

Developers and builders

You don't work for your consultants - they work for YOU. You pay them for the best possible designs, not to make their tasks easier. Reliance on a CAD software to 'automatically design' will cost you in terms of profitability and marketabiliy.

A 'quick and dirty' study often ends up permenantly built. Concentrate more on a well thought out plan from the best designer you can find. Understand that the Land Plan is essentially the 'developers' business plan. Ultimately the residents quality of living is in your hands.

Stewards of the Environment

For all 'treehuggers'; We can reduce the worlds infrastructure on newly developed land and redeveloped land an average of 25% by using the teachings in this book (combined with the methods taught within LandMentor). Environmental impact reduction is a byproduct of more advanced and efficient design. Reducing economic impacts for cities as well as increasing profitability for developers just takes little extra effort.

New 'green' trends can be just that, trendy. My own experience with green solutions have been mixed. My 'effortless - no mow lawn' has proven to be more expensive to maintain than if I had sod and a service to mow the lawn. Endorsing a solution just because it sounds right in theory, does not make it sustainable. If the proposed development yields more "green space" compared to developments of equal density, it's better for the environment, simple as that.

Elected officials

When we first discovered that planning innovation would reduce development costs, lower the cost of housing, lessen environmental impact, and increase safety, we shared this knowledge with our elected officials. We approached senators, representatives, and even leaders in the Metropolitan Council to see how they could help. Many (most) of those politicians were tied only to New Urbanism, especially those in control of Detroit's redevelopment - simply not interested! I went to Washington and met with the NAHB and EPA - not interested! Again, it's not New Urbanism.

It did not matter if they were Republican or Democrat. It seemed politicians only wanted a single solution and that any conflicting information would make it confusing to a voter. The only politician who made any effort was Senator Al Franken, and we thank him for his staff. Meanwhile I've been a keynote presenter at the Western States Planning Convention, the New York Professional Land Surveyors Conference, many Green and Sustainability Conferences, SLDI Best Practices Conferences, the California League of Cities Conference, The North Dakota AIA and the North Dakota Economic Development Conference, and many more, teaching Prefurbia.

Conclusion

You can make a difference. If everyone involved in the development process becomes more actively focused on the residents' needs above that of the developers and municipalities, we can make a positive change for future generations and the environment.

Those that have supported Prefurbia have all been passionate about the developments they build, approve, and engineer. The next time you see a cookie-cutter proposal, ask yourself how much passion and effort went into the design. In an era where technology has provided advancements in the past 50 years unimaginable in the previous 50 in every industry, we must ask - why has land development been stagnant?

Our cities and the hundreds of millions of people who dwell in them deserve much better.

Rick Harrison

Terms and Definitions

Some of the following terms and definitions are original, but many have been compiled from a variety of sources including Sustainable Land Development Inituative and Timothy Holveck of the Wisconsin Department of Natural Resources (www.dnr.state.wi.us).

Affordable Housing
Housing that has its mortgage, amortization, taxes, insurance, and condominium and association fees constituting no more than 30 percent of the gross household income per housing unit. If the unit is rental, then the rent and utilities constitute no more than 30 precent of the gross household income per rental unit.

Amenities
Features that add to the attractive appearance of a development, such as parks, recreational facilities, gathering places, and landscaping, and including the placement and location of utilities underground.

Arterial
A major street, which is normally controlled by traffic signs and signals, carrying a large volume of through traffic.

Architectural Blending
A form of design where the interiors spaces of a home or business becomes a function of the overall neighborhood design.

Architectural Shaping
Designing the perimeter of the home to take advantage of a non-rectangular lot.

BayHomes
An alternative to both conventional and neo-traditional communities. BayHomes use interconnected walkways and traditional architecture with enhanced interior and exterior views to create a warm, walkable neighborhood. BayHomes are single family detached townhomes with association-maintained common areas. Well adapted for production housing due to it's replicable footprint (building pad), BayHomes are staggered in a manner that creates unobstructed panoramic views from the kitchen and living areas looking outwards towards green open spaces from most of the units. The kitchen and living areas front onto common open space (typically not directly to a street) and the home's main vehicular entry is from the rear, via private drives, similar to homes that are serviced by alleys in traditional neighborhoods. All BayHomes have a front porch (facing open space) large enough for neighbors to congregate and relax. Each porch connects to a walk system that meanders through common areas. BayHome neighborhoods utilize approximately 50 percent less road infrastructure than traditional alley-based neighborhoods to construct.

Collector
A street designed to carry a moderate volume of traffic from local streets to arterial streets or from arterial streets to arterial streets.

Commercial zoning
A zoning area designated for community services, general business, interchange of services, and commercial recreation.

Common Open Space
Town squares, greens, parks, or green belts intended for the common use of residents.

Connective Neighborhood Design (CND)
A method of transitioning land uses that reverses the typical density progression (high density housing to low density) by doing the opposite. CNDs enhance neighborhood values by showcasing upper-end housing along all visible areas and positioning the higher density housing in a manner that preserves neighborhood values. A CND has the following design features:
- Land use transitions from the development's entrance to the rear, showcasing higher end housing (low density) first then transitioning to the lower priced homes (high density units) afterwards or in distributed pockets. An exception to this rule can be when the architectural controls guarantee that the lower priced homes will maintain or exceed the look and character of the higher-price-point homes.
- The neighborhood is interconnected with a walk system leading to defined destinations that provide active or passive recreation or commercial uses.
- Natural amenities are featured along the walks and, if possible, the street system, for all residents to enjoy.

Coving
Coving is an efficient method of land planning that utilizes a unique meandering road pattern, combined with an independently meandering home setback line, designed to vary the streetscape, thus adding visual interest. Coving also creates additional areas of open spaces along the street, referred to as 'Coves'. Density generally remains the same as a conventionally planned neighborhood, however, street function, flow, and safety improves and infrastructure reduced.

Environmental (or Economic) Density - ED
This is an accurate measurement that can be made to determine both economic and environmental impacts of a design. It was first made possible by the LandMentor technology.

Flow
The ability to enter and safely traverse the neighborhood with a minimum number of stops and turns while experiencing a sense of space and place. Flow can be applied to both pedestrian and vehicular systems. Flow significantly reduces time and energy in vehicular traffic.

Green
Refers to environmentally-friendly concepts, products, or services.

Growth
Growth is the natural and continued expansion of industry and population. Over the past 100 years, the average life span of an American citizen has increased from 48 years to 75 years. As a result, the population has increased accordingly. The population has grown from 76 million people to over 300 million people – or a 400 percent increase of people in the past century. In the same 100 years our population has survived two major world wars, while our industrial and intellectual-based products have gained world markets. All of this demands expansion for industrial, commercial and residential services. Of the 2,870,084 square miles in the continental United States' land mass, only approximately 156,000 square miles (approx. 100 million acres) of that surface is currently urbanized. With a population of 300 million today, this equates to an average density of 3 people per acre residing on only 5 percent of the surface area of the United States. See also "Sprawl".

Homeowner's Association
A nonprofit organization made up of property owners or residents who are then responsible for costs and upkeep of semiprivate community facilities and maintenance of common areas.

Infrastructure
Public utilities, facilities, and delivery systems such as sewers, streets, curbing, sidewalks, and other public services. Within the infrastructure may also be private services such as cable TV and internet.

LandMentor
A patented system combining technology, training and mentoring which expands upon the concepts, ideals, and examples within this book.

Mixed-use development
The practice of allowing more than one type of use in a building or set of buildings. In planning terms, this relates to the combination of residential, commercial, industrial, office, institutional, or other land uses within a specific and defined area.

New Urbanism
An American urban design movement that arose in the early 1980s. It is an approach to development that includes the reintegration of components such as housing, employment, retail, and public facilities into compact, pedestrian-oriented neighborhoods linked by mass transit. Public parks are encouraged, over private spaces. Also called "Neotraditional development," "Smart Growth," and "Traditional Neighborhood Design."

Open (Green) Spaces
A substantially undeveloped area, usually including environmental features such as water areas

Prefurbia
A neighborhood planning method that produces a preferred quality of life: low impact neighborhoods at densities similar to traditional neighborhoods but with more public and

private space and greater pedestrian connectivity, at significantly less development cost than traditional neighborhoods.

Rewards-based Ordinance
Also called incentive-based or bonus-based ordinances. This is a flexible zoning technique that permits a trade-off between the requirements of the land use regulation and the desired changes in those requirements by the developer. It allows for the relaxation of certain regulation minimums or other incentive, in exchange for an increased amenity that would benefit the residents of the development and their neighbors.

Right of Way (ROW)
A parcel of land that has a specific private owner, but some other party or the public at large has a legal right to traverse that land in some specified manner. The term likewise refers to the land subject to such a right. An easement is an example. Often a strip of land occupied by or intended to be occupied by a street, crosswalk, walkway, utility line, or other access.

Sense of Place
The constructed and natural landmarks and social and economic surroundings that cause someone to identify with a particular place or community.

Site Plan
A scaled plan, which accurately and completely shows the site boundaries, dimensions and locations of all buildings and structures, uses, and principal site development features (transportation and utilities) that are proposed for a specific piece of land. It may also include existing land contours, proposed elevation grades, walks, wetlands, ponds, and tree locations (existing and/or proposed). The site plan typically includes information such as total area of site, number of lots, type of units, and open space or park area if there is any.

Smart Growth
An urban planning and transportation theory that concentrates growth in the city for the purpose of growth management. It's policies advocate compact, transit-oriented, walkable, bicycle-friendly land use, including neighborhood schools, streets, and mixed-use development with a range of housing choices. See also "New Urbanism" and "Neotraditional development.

Sprawl
Inefficient use of developed land, because of either wasteful site plan designs or municipal regulations that result in wasteful use of the land – or both of those combined. See Growth.

We exclude from the definition reference to unplanned, uncontrolled spreading of development, as growth is the natural continued expansion of industry and population, and people are free to exercise their constitutional right to choose where to live. In this book, when we write about reducing sprawl, we are referring to definition "A," wasteful site design that results in wasteful utilization of land and natural resources.

Sustainable Land Development
The art and science of planning, financing, regulating, designing, managing, constructing and marketing the conversion of land to other uses through team-oriented, multi-disciplinary approaches which balance the needs of environment, economics, and existence – for today, and future generations.

Sustainability
A characteristic of a process or state that can be maintained at a certain level indefinitely.

Tax Increment Financing (TIF)
A tool to use future gains in taxes to finance the current improvements that will create those gains. Commonly used for redevelopment and community improvement projects, specifically public projects such as a road, school, or hazardous waste cleanup.

Traditional Neighborhood Design (TND)
A compact, mixed-use neighborhood, where residential housing often has detached garages, accessed via alleys, and commercial, schools and civic buildings are within a close proximity. The TND is most often associated with a grid street pattern and blocks that are much shorter in length than those typically associated with suburban design. The TND is often referred to as New Urbanism. See also "Neotraditional development" and "New Urbanism."

Urban Planner
Often also known as a city planner, an urban planner formulates plans for the short and long-term growth of a city. They study land use compatibility, economic, environmental, and social trends. In developing their plan for a community, urban planners consider a wide array of issues such as air pollution, traffic congestion, crime, land values, legislation and zoning codes. They focus on the macro issues in planning. They may monitor or control the ordinances and regulations of a municipality as well as be a bridge between the city and the land developer and land planner. Often, the urban planner represents the interests on the regulatory side of town planning instead of the design side. However, like the civil engineer and land surveyor, the urban planner may also act as the land planner and design site plans. While the urban planner may not have the engineering or surveying technical understanding of the other two professions, they are likely to have the aesthetic talents to apply to the land planning the technical professions may lack.

Variance
A requested deviation from the set of rules a municipality applies to land use known as a zoning ordinance, building code or municipal code.

Zero Lot Line
The location of a building in such a manner that one or more of its sides rests directly on its lot line.

Neighborhood Showcase

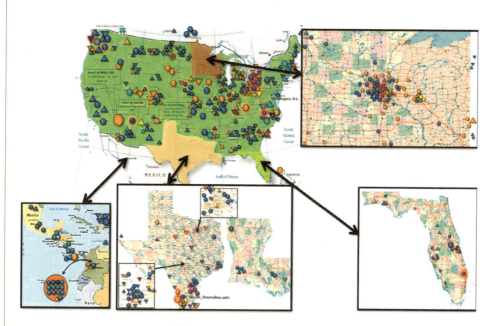

Examples of Prefurbia

The above map represents just a few of the 1,100+ neighborhoods we designed in 47 States and 18 Countries using the Prefurbia methods introduced in this book.

The following showcased neighborhoods are under construction or approved at time of publishing unless otherwise noted. Within this 5th edition they represent the very latest innovations, methods, techniques, and technological advancements.

Unlike other products that can be brought to market quickly, land development can take many years to go from an approved plan to a finished and established neighborhood, sometimes decades.

All examples were designed using the LandMentor System.

For more examples, please visit our web site at www.rhsdplanning.com

TRES LAGOS

McAllen, Texas – Rhodes Enterprises, Inc.

Tres Lagos (Three Lakes) the largest Master Planned Community in the Rio Grande Valley, consists of over 2,571 acres and will ultimately have approximately 8,000 homes. This mixed use community is situated on the northern portion of McAllen, Texas. The original 'Master Plan' that was submitted used conventional methods that most large scale communities are designed. The major internal boulevards with 150' wide right-of-way's were reduced from the original plan (shown below) by over 8 miles - resulting in a gain of over 76 useable acres of land!

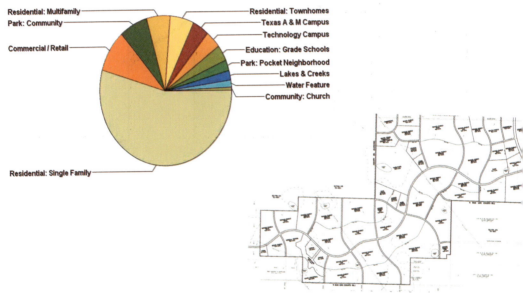

Original Tres Lagos Master Plan

A Neighborhood Marketplace, Estate Homes, Parks and Lakes define the main entrance.

Project Showcase 181

ELEONOR

Tomsk, Siberia (Russia)

This beautiful neighborhood is the first phase of a very large Master Plan (done previously by another consultant) located southwest of the town of Tomsk. Coving was used to increase efficiency and home value compared to the 'before' plan shown in the inset, resulting in a 13% increase in density with a 19% decrease of infrastructure. The center townhome grid, as well as the 'trendy' multiple round-abouts were existing so they could not be revised.

In addition to gaining density and efficiency, Prefurbia increased connectivity, function, safety, flow, premium locations, and open space.

Russian suburban design is very similar to the American suburbs in form and function. The site is bordered by a beautiful creek and the design reflects a reduction of impact on slopes. This neighborhood is under construction, the 3D LandMentor rendering shows Russian architecture.

Area	Approx. 300 Acres
Coved Single Family Lots	521
Average Lot Size	17,705 sq.ft
Area Meandering Front Yard	75.1 Acres

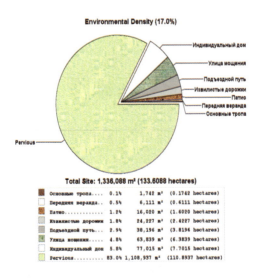

Project Showcase 183

BEFORE PLAN

ZIPA QUIRA

Near Bogota, Colombia by AMARILO

This unique neighborhood contains nearly 900 single family homes. It harnesses every new planning method of Prefurbia to create a truely 21st Century Master Planned community.

The original plan consisted of many short segments resulting in terrible 'flow' of traffic. It was just natural for any developer to assume the original plan would provide the most density. Our alternate plan achieved the same density with 1/3rd less street, while eliminating monotony and providing a street pattern that would require less travel time and energy. In addition, walking connectivity was provided where none existed before. We designed a similar Master Plan in Panama estimated at 10,000 new homes, under construction. This neighborhood is being built as shown.

Area	248.23 Hectares
Single Family Coved Lots	848
Average Lot Size	1,227 m2
Minimum Lot Size	900 m2.

The Master Plan (massing of buildings) showing the terrain, streets, and main trails.

Project Showcase

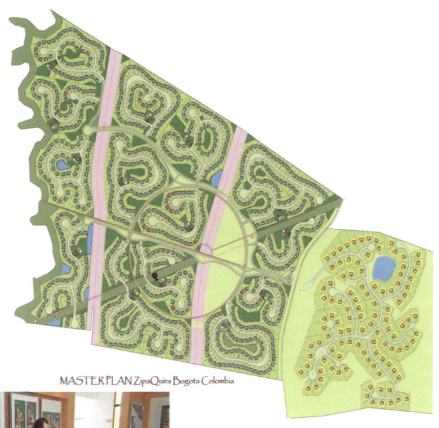

MASTER PLAN ZipaQuira Bogota Colombia

Clockwise from above:

The Master Plan, An interior courtyard of one of AMARILO single family homes, The main street and trail patterns and Adrienne Harrison at the home sales office of a nearby neighborhood that was designed by Rick Harrison Site Design near Bogata.

PRAIRIE CROSSING

Kildeer, Illinois – Pulte Homes

Almost all of the developments featured in this book are from the lower to upper middle class strata of the marketplace. This is because very upscale development will look good no matter how the site plan is designed with expensive homes, landscaping, and cars parked in sight. That stated, the methods introduced in this book as applied to upscale housing does make a major difference. Prairie Crossing was to be Pulte's most luxurious development. Access to this luxury 75.1 acre Kildeer (NW suburb of Chicago) neighborhood is along the site's south border, from West Cuba Rd, then turning north, into North Prairie Lane.

Kildeer had a 'cluster ordinance', which allows ½ acre lots, with a 50% open space requirement. Coving increased the lot size from the minimum 21,780 sq. ft., to an average size of 25,544 sq. ft, while still achieving the allowed density goal.

Although North Prairie Lane appears to be an exceptionally long cul-de-sac, it was allowed due to the addition of an almost hidden landscaped emergency access, providing access via N Quentin Road, the main street along the west border of the property.

Coving was used to enhance the feeling of space, while ensuring all 54 homes had magnificent views of the site's many natural and manmade features, while also minimizing development impact.

A beautiful meandering natural white stone walking trail connects stolling neighbors both through the site and around the perimeter. Its placement provides maximum exposure to the creek, aquatic garden, landscaped trellises and relaxation areas.

The city required 50% open space on this site; we achieved 65% open space (49.1 acres).

This neighborhood is fully constructed, Google Maps image shown.

Area	75.1 Acres
Coved Single Family Lots	54
Homes per Acre	0.72
Average Lot Size	25,544 sq.ft
Minimum Lot Size	21,780 sq.ft.
Open Space	34.9 Acres
Area Meandering Front Yard	14.2 Acres
Total Openness	49.1 Acres (65% of Site)
Area Street Public Paving	3.96 Acres

REMINGTON COVES

Otsego, Minnesota

The free-flowing organic design was the first implementation of coving where homes formed 'blocks within blocks' creating internal open spaces that greatly expanded viewsheds. This method of coving creates a rural feel while achieving an impressive overall 3.05 homes per acre.

The city's minimum lot size was 9,000 sq. ft; yet this pioneering design yielded lots averaging almost 1/3 larger in area, 12,750 sq. ft. on this 37-acre neighborhood. This neighborhood stalled at the recession, like all outer ring suburban developments in Minnesota. As the recovery began, the single family lots were picked up by Ryland Homes, one of the national home builders with very strong sales - selling out in just a few months!

Reasonably priced housing choices, walking paths with multiple routes leading to visible green spaces, a 1.2 acre park, two gazebo's and fully equipped children's playground enhance livability. Two wide paths double as emergency accesses, adding north / south connectivity that places Remington Cove's walkability on par, or better, than traditional or New Urban neighborhoods.

Area	37 Acres (net)
Total Housing Units	113
Coved Single Family Lots	70
Number of BayHomes	43
Homes per Acre	3.05
Average Lot Size	12,751 sq.ft
Minimum Lot Size	9,000 sq.ft.
Park Area	1.2 Acres
Area Meandering Front Yard	9.3 Acres

SETTLER'S GLEN

Stillwater, Minnesota – Lennar Homes

This 125 acre neighborhood is located in Stillwater, a quaint rural town located along the St. Croix River, and a popular tourist destination for antique shopping.

Coving was used to minimize earthwork and preserve many trees, wetlands and environmentally sensitive areas that surround the site, which includes a trout stream, known as Brown's Creek. The site plan has 220 single family homes, set on a minimum lot size of 7,000 sq. ft. The use of Coving increased the average lot size to 10,299 sq. ft.

In a July 10, 2005 article in the *San Diego Union-Tribune*, Bob Swanick, regional vice president of US Home (now Lennar, but was Orrin Thompson Homes at the time), commented, *"We were looking at communities of narrow home sites, so it was important for us to end up with a community that had a great deal of curb appeal with a maximum amount of parks and still give us the density we needed. Not only did he [Rick Harrison] improve our density, he lowered our development costs, and that allowed us to be a little more affordable than the competition."*

Meandering trails connect homes to each other and beautiful views of the wetlands, stream, three parks, and athletic fields. Total openness, including the park-like greens along the streetscape, is 37% of the site.

When we first presented the plan we were informed that only 40% of the home front could be garage face (to prevent overload of garage door views) by the ordinance. We demonstrated that with inventive architecture homes could have a forward access garage(s), 3 car garages, and a front porch for character. For this demonstration the City gave us a variance due to the successful architectural alternatives. This neighborhood is fully built and sold strong - even during the recession!

Area	125 Acres
Total Housing Units	220
Homes per Acre	2.5
Coved Single Family Lots	220
Average Lot Size	10,299 sq.ft
Minimum Lot Size	7,000 sq.ft.
Open Spaces	67.5 Acres
Area Meandering Front Yard	6.3 Acres
Total Openness	37 Acres (37% of Site)

THE CHANDLER

Frankfort, Tennessee - Davenport Development

This beautiful neighborhood was designed under the guidance and passion of C. Michael Davenport, named after his daughter, Chandler. He previously developed the site as a Christian retreat - with a wooden bridge, a winding street, and a heart-shaped cul-de-sac around a large gazebo/150' tall flag pole in its center.

As unique as the vision of Mr. Davenport, was the fun we had working under his direction to enhance it with the methods taught in this book. Prefurbia opened his eyes to new ways to build and develop. All the elements you learned about, including the Home/Business land use to conform to the 'mixed-use' zoning have been used on 'The Chandler'. The level of detail was incredible - including his desire to replicate the architecture of the original 'Sears Homes' a century ago.

Area	19.6 Acres
Units per Acre	4.6 du/ac
Single Family Lots	40
Mulit-Family Units	50
Total Units	90
Total Common Space	6.1 Acres

The genius of Mr. Rick Harrison allowed The Chandler to be a very special, unique, creative, environmentally friendly, and a out of the box non-cookie cutter development. After talking with other land planners across the country I made the decision to go with Mr. Harrison's Prefurbia. The entire process was very professional. His responsiveness was critical to our timeline. I am so pleased and proud of The Chandler. It was a pleasure to work with Mr. Rick Harrison.

Sincerely; C. Michael Davenport

Project Showcase 193

GREENBRIAR HILLS

Buffalo, Minnesota - TSM Greenbriar LLC

Greenbriar Hills will redefine value in the midwest housing market using the full menu of ingredients in the Prefurbia recipe. This architecturally controlled neighborhood must compete with the nearby major builders who most often build in the standard suburban cookie-cutter pattern, with the exception of Pulte who has three new coved communities in the region. In most markets, the big builders closely mimic each others products. Greenbriar Hills offers a better neighborhood than the competition with park-like settings and water features everywhere, but will also push the envelope of housing using 'architectural shaping', 'blending', and NextGen BayHomes to provide a significant advantage over the offerings of other builders competing at similar price points. As with everything presented in this book, the more developers and builders can abandon the past to serve the residents better, the more likely they will be competitive. Greenbriar Hills (under construction as of the writing of this book) will show consumers what is possible. That's how we change the markets.

Area	99.6 Acres
BayHomes	64
Coved Single Family Lots	152

"As a newcomer to the land development process we were looking for a way to provide the homeowners and the community with something special and unique. "Perfurbia" provided a guide to enhanced livability that would capture the beauty of Minnesota. From understanding the distinction of the recreational aspect of the development with the enhancements and value provided to the homeowner "Perfurbia" is the blueprint for the new standard of architectural design and sustainability."

Chris Trapp TSM Capital, LLC

Project Showcase

GREENBRIAR HILLS
of Buffalo, Minnesota
Development Stage PUD
by: TSM Greenbrier, LLC

PASEO DE ESTRELLA

Albuquerque, New Mexico – D.R. Horton

This 40 acre neighborhood is located in the master planned community of Vista del Norte. To get there, take State Hwy 25, exit West onto Osuna Rd NE (Hwy 25) and turn into Sidewinder Drive.

This award-winning neighborhood was New Mexico's fastest selling community in 2004. Paseo de Estrella *"was so well received, we sold all our lots within 12 months,"* says Bob Prewitt, a former vice president of D.R. Horton's New Mexico division who stated; *"Harrison's plans provided us with unique communities that return greater profits and market advantages over our competitors."*

Through its organic design this efficiently-planned 162 home neighborhood blends Albuquerque's unique desertscape into the lives and homes of its residents.

The neighborhood includes homes for first time homebuyers, as well as move up markets. Usable land area was increased by the coved approach that consumed much less land in ROW, in fact 39% less linear ft. of street – from 7,500 ft. down to 4,600 ft. The plan's efficiency freed up enough additional land to add more parks, a beautiful meandering walk system, and lots that averaged 20% larger than competing subdivisions. Average lot size was 7,814 square feet yet density is four homes per acre.

Homes were positioned at graduated angles and varying setbacks along a meandering boulevard that when combined, offer greater scale and create a striking streetscape.

Twenty percent of the net landscaping was water-demanding species and was reserved for the 5.2 acres of common park areas; the remaining 80% of landscaping was xeriscaped, to blend the region's natural surroundings into resident's tangible environment.

Impervious surface area comprises 39% of the site.

Paseo de Estrella is fully built.

Area	40 Acres
Homes per Acre	4.0
Coved Single Family Lots	162
Average Lot Size	7,814 sq.ft
Parks	5.2 Acres

FRIEDËN

Fredricksburg, Texas

Friedën has been a labor of love for me – it took three years to complete design and entitlements for this mixed-use development. I'd spent decades spent looking for the right location, in the right market, at the right time for this kind of unique project.

Fredericksburg is in the Texas Hill Country, north of San Antonio and west of Austin, and has a distinctly German heritage in terms of history, culture and architecture. Many of the older homes in Fredericksburg's Historical District are old farmhouses and Sunday Houses – most of them use native limestone rock for walls and gabled metal roofs in a style unique to the German communities in the Texas Hill Country. The Modern Farmhouse style chosen for the homes to be built in Friedën are reminiscent of these old, historical structures.

I have had the opportunity to see quite a few of Rick's completed designs – both as a developer and as a development consultant – and I've observed that his projects tend to look better as they age. This is significant, because one of the dirty little secrets in residential development is that most neighborhoods that utilize grid patterns don't age particularly well.

The Modern Farmhouse style allows for consistency in the community, but enough architectural flexibility to be distinctive. And Rick Harrison's coved design provides several different lot configurations, miles of connected walking trails and open green space behind all the homes. It is a unique combination – one unique enough to have been selected as the first Southern Living Inspired Community in Texas...

...Skip Preble, Land Analytics and developer of Friedën

Area	219.5 Acres
Area of Comm / School/ Multi-Family	72.9 Acres
Coved Single Family Lots	228
Average Lot Size	13,264 sq.ft.
Parks & Street Islands	43.1 Acres

Project Showcase 199

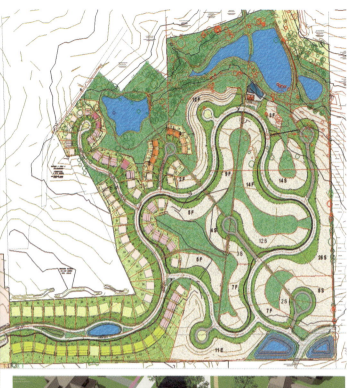

SILVER PONDS

Kreps Development - Rogers, Minnesota

We wrote about trends that come and go, leaving behind housing or developments that are either dysfunctional, not in style, or obsolete. About 25 years ago the trend of 'Ghost Platting' was a highly promoted alternative for non-sewered large lot development where the large lots were pre-subdivided for the expected eventuality of public sewer reaching the site. Planners (including us) simply placed smaller lots - often in unbuildable areas to meet the letter of the law, but few took it seriously. The main problem was that if someone had a large lot they also wanted the freedom to place the home exactly where they wanted it - rendering future 'ghosted' lot useless. Another problem as small lots would have an odd look when sewer was available - small homes would intermix with the large estate homes. Well, as trends that go, often come around again. The 'Ghost Platting' has recently being promoted by the Metropolitan Council (twin cities) growth plan.

   This will be the first 'Ghost Plat' in the region. It is intended to serve as a model for others to emulate. The current and near future single family lot width in this region is between 60 to 70 feet wide. Only a few areas of the site could support septic fields. We designed around the existing homes, wetlands, and proposed septic areas, then platted 1 acre minimum lots with the required 150' minimum frontage along the street right-of-way. At the same time creating a 'coved' streetscape that would work with 70' wide lots with a specific location for the first home on the lot. When public sewer is available within a decade, the owner of that lot can decide to sell the 'ghost lots' which will already be platted to help pay for the assessments that will be charged to the home owner, or just keep on a large lot which will have extra value to the next home buyer.

Perhaps we can make this 'trend' stick. Currently as of this writing, the city gave sketch plan approval, and is on the way to preliminary plat.

Note that the term 'Ghost Plat' is also known as 'Shadow Platting'.

1" = 100 feet

Before Public Sewer

After Public Sewer

ROSEHEART

San Antonio, Texas – Sitterle Homes

Roseheart is an 83.5 acre active adult neighborhood in San Antonio, located off of Bulverde Road & Roseheart Road.

Roseheart has a village atmosphere with intimate spatial relationships & multiple interior green spaces containing water features, pool, tennis courts, gazebos, mini-parks, walking trails and a clubhouse.

The vision for this 241-home community, was to create a plan that would enable the maximum number of residential lots abutting a private greenbelt perimeter and heavily wooded interior while preserving most of the existing abundant trees. The green belt is covered in mature native trees (oaks, elms, and wild persimmons), and includes two wooded creeks & a large cave preserve, providing residents access into a secluded refuge from city life. These natural constraints also rendered 20 acres unbuildable; however, these same features were then turned into an advantage by utilizing the area for natural drainage and detention.

As a result of coving, streets lengths were reduced substantially, so there was fewer sewers, fewer waterlines and less paving, resulting in a savings of close to $200,000 in development costs.

Frank Sitterle, a 41-year Texas home-building veteran who has won numerous awards, is the site's developer. *"Everybody loved it,"* he said. *"Every single lot is a greenbelt lot. This is a huge marketing advantage, and we were able to really maximize our return because we could get a premium for every lot we had."* Density is 2.73 homes per acre, with an average lot size of 9,122 sq. ft.

Impervious surface area comprises 27.41 acres, or 32% of the site.

The site is entirely sold out.

Area	83.5 Acres
Homes per Acre	2.73
Coved Single Family Lots	241
Average Lot Size	9,122 sq.ft
Parks	22.1 Acres
Area Meandering Front Yard	16.0 Acres
Total Openness	38.1 Acres (46% of Site)
Area Street Public Paving	6.31 Acres
Impervious Surface Area	27.41 Acres (32% of Site)

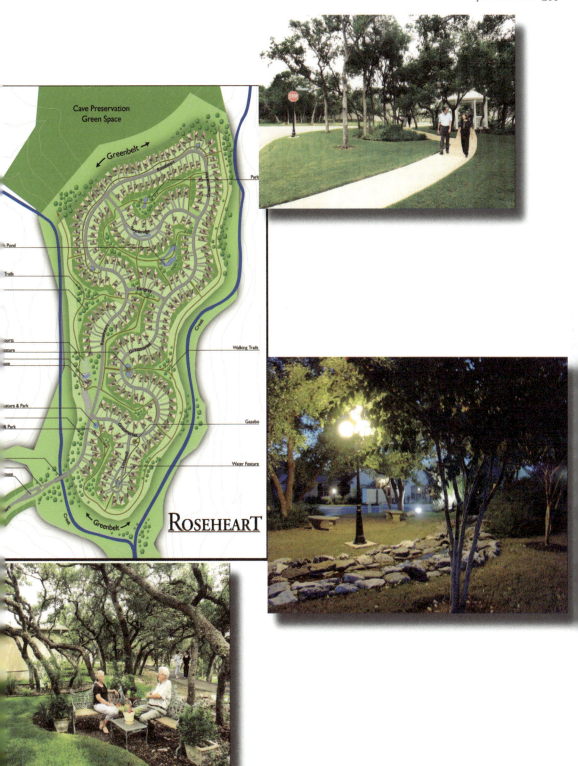

TRASONA AT VIERA

Viera, Florida - Viera Development & Viera Builders

Viera is a large scale Master Planned community near Melbourne, Florida. There were several TND based studies done before Prefurbia was chosen. One example is shown above left of the Prefurbia plan for comparison of the first 700 lot section below. We were introduced to the developer by Hassan Kammal of BSE Consultants who engineered a neighborhood we had previously planned in the region. The Viera Company has a development and home building division.

There were certain 'smart growth' elements that were required and incorporated in the Prefurbia design. The density and housing mix of the TND was also retained. This is the world's first development where the homes were redesigned to make maximum use of the architectural shaping and blending. Viera Builders replaced the original rectangular floor plans (see page 144).

Because of previous planning criteria, the site is bisected into four quadrants which would typically create separate neighborhoods, however, in Trasona, a circular main trail binds neighbors together. An additional straight tree-lined main trail becomes another neighborhood feature. Elegant meandering walks link residents to the main trail. Traffic diffusers help create neighborhood character and sense of place while increasing both vehicular and pedestrian safety. Ponding area required was 18% of the land area. Street length reduction was an outstanding 38%!

	Before	After
Site Area	292.3 Acres	289.7 Acres
Average Lot Size	8,146 sq.ft.	9,877 sq.ft
Area Public Street Per Home	2,143 sq.ft.	1,707 sq.ft.
Total Length of Public Street	44,070 lin.ft.	27,263 lin.ft.
Number of Lots	710	703
Number of Street Intersections	54	18

BEFORE PLAN

Project Showcase 205

Biography

Rick Harrison

Growing up in Oak Park just outside Detroit, I dreamed of one day designing cars. At the age of 15, my mother, Carol Minowitz, came home and said, I've arranged a job interview for you. After showing a few car sketches to Don Geake, of the land planning firm Geake & Associates in Southfield, Michigan, I was hired to begin an after school job in land planning as a draftsman. Within a few weeks, Don had me laying out subdivisions. I was designing land developments before I was old enough to drive!

At Geake & Associates, we used marker pens to create freehand designs. We saw ourselves as artists. At no time was I taught how much development costs to construct, or how to reduce infrastructure. If adding an extra bridge on a site would increase density, I threw that in, never thinking the cost of a bridge would be more than gaining a few lots.

We kept developers happy by improving density. One way (trick) to maximize density was to 'trace' a site survey larger than it actually was. This created the ability to show a larger number of houses on the site. The mis-represented density vanished when the engineer or surveyor tried to make the site plan work. Given an unworkable plan, the engineers and surveyors often changed our designs to the point that they no longer looked anything like what we initially drew. The bigger problem was that the developer estimated costs and profit based our original inaccurate information.

During the six years working for Don's planning office, it was never suggested that we consider the cost of constructing the streets, sewers, or drainage systems. Not once did we refer to topographic maps to determine the best position for a street or home, to minimize the impact of earthwork. That was the job of the engineers, not us.

Unfortunately, little has changed in the land planning field in the half century since I worked for Don Geake. Developers still assume their planner has taken engineering and surveying issues into account, but in most cases this is not the case.

By 1974, the auto industry was in crisis due to the oil shortage. A stalled auto industry in Detroit meant there was little new development work. However, at the same time I was approached with a new opportunity: to become a Land Developer.

At age 21, I was in business with my stepfather Abe Minowitz trying to correct some (many) construction mistakes that had been made constructing a 64-unit development he had invested in, called Robert Arms in Newburgh, New York. I was excited about the opportunity.

When I arrived at Robert Arms, I had to solve what was essentially an earth-moving problem. The soil under one of the buildings had eroded so badly that the bottom of one of the basement foundation was visible and there was a 90-foot drop to the river valley below. The building was in danger of sliding away!

Abe asked me how much it would cost to fix the problem. I, being an ignorant 21-year-old ex-land planner said, *"It will only be $600 a day for the bulldozer -- plus whatever* *dirt costs.*" When we got the bill, the "*whatever dirt*" cost over $10,000 – the equivalent of about $42,000 today after inflation. This was for just one apartment building.

I received a justified stern lecture from Abe. This experience taught me that a planner must look at more issues than just the almighty yield. In fact, I realized that, to be effective, a planner must have a full understanding of surveying and civil engineering.

After solving Robert Arms problems, I relocated to Texas (where the jobs were in the mid 70's) and began as an apprentice under the supervision of Paul Lederer, and Chalmers Miller who were land surveyors and civil engineers. In addition to learning as much about surveying and engineering as I could, a new hobby began – developing computer programming related to my work.

Eventually, I left Houston for Dallas to become Head of Planning for Herman Blum Engineers, one of the largest consulting firms in Texas. It was there that my programming hobby evolved into a marketable civil engineering and land surveying software business.

After selling 20 'tapes' of my first software in the Dallas market, eventually, I got a call from Hewlett Packard about a soon to be launched desktop computer called the HP-87. They wanted me to write a surveying package they could promote with their computer. A very succesful 20 year collaboration with HP began.

In 1982, I developed my first *Site Computation & Design* software package. In a time when others sold their software for upwards of $20,000… (back then, software did not do much) I estimated if I priced the software at $895, undercutting everyone on the market, the phone should ring off the wall. After a national advertising effort, it did. Within weeks, I left my land planning career behind to become a software developer.

Over the next two decades, we sold about $20 million in systems. Because the equipment was so bulky (about 200 pounds) I began flying lessons. I soon owned and flew a Cherokee 6, upgraded to a Mooney, a P-210 and then a Piper Malibu. With over 4,500 flight hours servicing customers all around the country I gained an intimate 'pilots view' of the development patterns that were taking place. One thing that was very clear - all this new software technology made planning worse because of repetition which became the norm.

After a few years, I moved on to a new challenge: use what I learned in the sofwtare business, to help improve the methods of land that was being planned. That is how Rick Harrison Site Design Studio was founded. I soon discovered *Coving*, and with it success came faster than I ever imagined. We have been continually discovering new ways to raise development standards in both design and regulation.

With the land planning business flourishing, we decided to teach others the knowledge that we have gained, so that these methods, concepts, and even technologies can improve the quality and sustainability of all future development. We decided to write a book. In 2004, we began writing this book and a few years after that began developing an entirely new form of software, along with comprehensive design training, now sold as LandMentor. LandMentor is a patented software technology created specifically for sustainable land development design. We did not do this alone, mind you...

Skip Preble, a financial consultant for land developers helped us design analytical aspects of LandMentor financial modeling. Skip, president of Land Analytics, LLC (www.landanalytics.com), sums up Prefurbia:

Everyone in the land development process should devour this book, because the vast majority of Americans want what it describes. Prefurbia describes a more livable, attractive development environment – one that minimizes pavement, allows for "walkable" neighborhoods, and provides streetscapes that will get more attractive as they age. In my land development consulting practice, I have had the opportunity to test these methods on a number of occasions, and in every case the resulting designs have not only produced attractive, livable communities, but they also reduced development costs and accelerated absorption rates compared to competitive developments using conventional design. Prefurbia is absolutely a win-win for all involved.

Many other consultants have been instrumental in overcoming objections of engineers who fight progress and change. They represent the future, a select leading edge group of consultants who believe in a brighter and sustainable future and have proven instrumental in making change possible.

When I began writing software, my fellow workers at Herman Blum Engineers snickered and said 'he will never sell any'.

If only I had a nickel for every naysayer who said it won't work.

We designed over 1,100 developments in 47 States and 18 Countries.

For more information and the latest on the concepts presented in this book, you can go to the Rick Harrison Site Design website at *www.RHSDplanning.com*.

To continue your education on Prefurbia visit our LandMentor website at
www.land-mentor.com

Photo Credits

Photo Identifiers and Image Credits

Unless indicated below, all photos were taken by Rick Harrison and graphics were created by Rick Harrison Site Design and are copyrighted by Rick Harrison. Only images not identified within the pages of the book are listed here.

Cover:	Jamil Ford & Rick Harrison.
Page 1:	Toronto, Ontario, Canada.
Page 3:	Mesquite, Texas.
Page 16:	© The New Yorker, used by the author with permission.
Page 19:	St. Michael, Minnesota.
Page 21:	Figure 2.3: Doug DeHaan of Homescape Design Group.
Page 32:	Placitas de La Paz & Conventional Subdivision Coachella, California.
Page 41:	Middleton Hills, Middleton, Wisconsin.
Page 43:	Park Place TND: Farmington, Minnesota.
Page 57:	Westridge Hills, Eau Claire, Wisconsin.
Page 58:	© Sustainable Land Development International, used with permission.
Page 65:	Subdivision in Plano, Texas.
Page 67:	Ramsey Town Center, Ramsey, Minnesota.
Page 73:	© The Oakland Companies, used with permission.
Page 81:	Suburban Development in West Des Moines, Iowa.
Page 87:	Villages at Creekside in Sauk Rapids, Minnesota
Page 102:	Original street geometry plan by Lloyd and Tryk Architects provided to Rick Harrison Site Design by the project developer.
Page 107:	Commercial strip center with senior townhomes behind it in St Michael, Minnesota.
Page 127:	Hamilton Mills, Georgia.
Page 129:	Remington Coves - Otsego, Minnesota taken from BayHome porch.
Page 139:	Streetscape in Roseheart, San Antonio, Texas.